IMAGES of America

Historic Bridges of Milam County

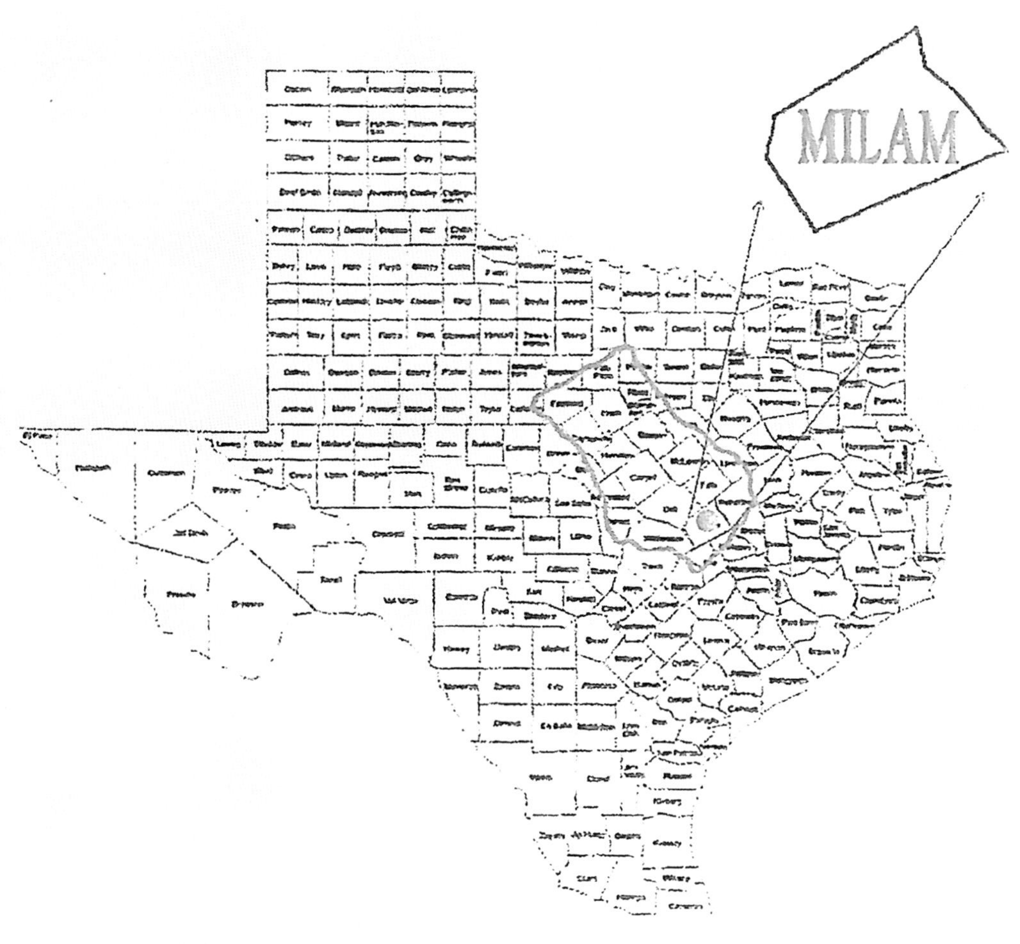

Milam County is located in East Central Texas, approximately 150 miles inland from the Gulf of Mexico. Cameron is the county seat; other towns in the county include Rockdale, Milano, Gause, Buckholts, and Thorndale. Through its years of existence, Milam County has constructed a number of bridges to cross major waterways in the county, including the Brazos, Little River, and San Gabriel Rivers. (Courtesy of the Milam County Historical Commission.)

On the Cover: This picture of Worley Bridge was taken by David Galbreath in June 2014 at the rededication ceremony for the restored bridge. (Courtesy of David Galbreath.)

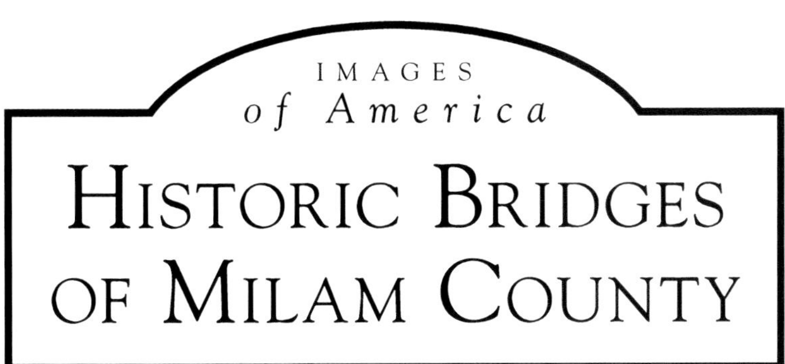

David Galbreath, Carolyn Temple,
Lucile Estell, and Joy Graham

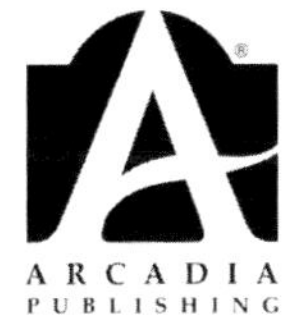

ISBN 978-1-4671-2491-1

Published by Arcadia Publishing
Charleston, South Carolina

Printed in the United States of America

Library of Congress Control Number: 2016959854

For all general information, please contact Arcadia Publishing:
Telephone 843-853-2070
Fax 843-853-0044
E-mail sales@arcadiapublishing.com
For customer service and orders:
Toll-Free 1-888-313-2665

Visit us on the Internet at www.arcadiapublishing.com

This book is dedicated to the members of the Milam County Commissioners Court, past and present, for their unfailing support of historic preservation in Milam County.

Contents

ACKNOWLEDGMENTS

This book has four authors, but their work does not account for the sum total of the book. For a number of years, ever since his retirement from the US Air Force, David Galbreath has studied these bridges. He has documented his research and generously shared it with many.

The Milam County Historical Commission has untiringly supported this effort. Its contribution includes actually selling the finished product.

Jack and Beth Brooks, members of the Milam County Historical Commission, have shared many of their photographs.

Tense Tumlinson of Cameron has shared information compiled by her late husband, Jack Tumlinson, who was the first person to inventory the Milam county bridges.

Tracy Galbreath-Young, daughter of David Galbreath, has helped with proofreading and other technical issues.

Charles King, curator of the Milam County Museum, provided valuable assistance in locating and preparing photographs for reproduction.

Jessica Bowling provided technical assistance as we started our project.

George and Eleanor Schmidt Hollas of Cameron shared information complied by Eleanor's late father, Victor R. Schmidt, who was senior resident engineer with the state highway department that supervised the construction by Thomas & Ratliff of Rogers on the Little River Bridge.

Darlen Graham has shared her organizational skills and talent for details in organizing the book. She has contributed many hours to the development of this publication.

Dolores Mode of Cameron collected many photographs and shared information gained from many years of service on the Milam County Historical Commission.

INTRODUCTION

The concept of a man-made device to enable the public to move around natural obstacles is not a new one. The name for these early devices goes back to the early cultures of the Greeks and Romans and to the Mayans, who were long admired for their bridges and other structures of beauty. There is absolutely no connection between these early structures and the popular game by the same name.

Milam County presented few natural obstacles to free movement. In the beginning, the county was approximately six times as large as it is today.

The history of Milam County began in 1836 when Robert Leftwich obtained a colonization grant from Mexico. Boundaries of the grant followed the Navasota River, turned southwest along the Old San Antonio Road (now a part of the El Camino Real National Historic Trail), next to the divide between the Brazos and Colorado Rivers, then northwest to the Comanche Trail, and back to the Navasota River. This area was about one-sixth of the land area of Texas.

Reductions have brought it to its current size. The bridges discussed in this book are in present-day Milam County.

Milam County today is one of the loveliest counties in the state. Its pleasant terrain, however, did call for bridges to be built to overcome obstacles, both for transporting people and goods.

Consider the stunning, impressive face of Milam County. Three major rivers are found there, including the Brazos, which is called Brazos de los Dios (the arms of God) by the Spanish because of its pervasive wandering nature.

The Little River is a river in Central Texas in the Brazos River watershed. It is formed by the confluence of the Leon River and the Lampasas River near Little River in Bell County. It flows generally southwest for 75 miles until it empties into the Brazos River about five miles southwest of Hearne at a site called Port Sullivan in Milam County. The Little River has a third tributary, the San Gabriel River, which joins the Little River about eight miles north of Rockdale and five miles southwest on Cameron.

The Little River has had several names. In 1716, Domingo Ramon reached the river, and he named it the San Andres. When the Marquis de San Miguel de Aguayo found the river in 1719, he named it Espíritu Santo because he came upon it on the eve of Pentecost. Pedro de Rivera y Villalon found the river in 1727 and believed it was simply the arm of the Brazos. The name *San Andres* was generally used during the colonial period; however, in the early years of the Republic of Texas, the river was called the Little River.

The face of Milam County also includes creeks, including Brushy Creek, Yegua Creek, Pond Creek, Big Elm Creek, Cedar Creek, Donahue Creek, Sandy Creek, and other small creeks.

The presence of these rivers and creeks necessitated and encouraged the placing of bridges.

In 1994, the Milam County Historical Commission, headed by Bonnie Belle Holder of Cameron, realized and recognized the importance of these magnificent structures. The commissioners asked Jack Tumlinson, a Cameron resident, to conduct an inventory of these significant bridges. A historical inventory of the bridges was intended to encourage a preservation plan for these faltering giants. A map with a suggested tour route was published by the Milam County Historical Commission. Four of the bridges were named on the map: Elm Creek, Little River, San Gabriel, and Brushy Creek. Tumlinson took pictures of each identified bridge and donated them to the Milam County Historical Museum in Cameron. In addition to the four bridges named on the map, the inventory included the following: Galbreaths Crossing, Bryant Station, Donahoe, San Gabriel, Marlow, and Faubion.

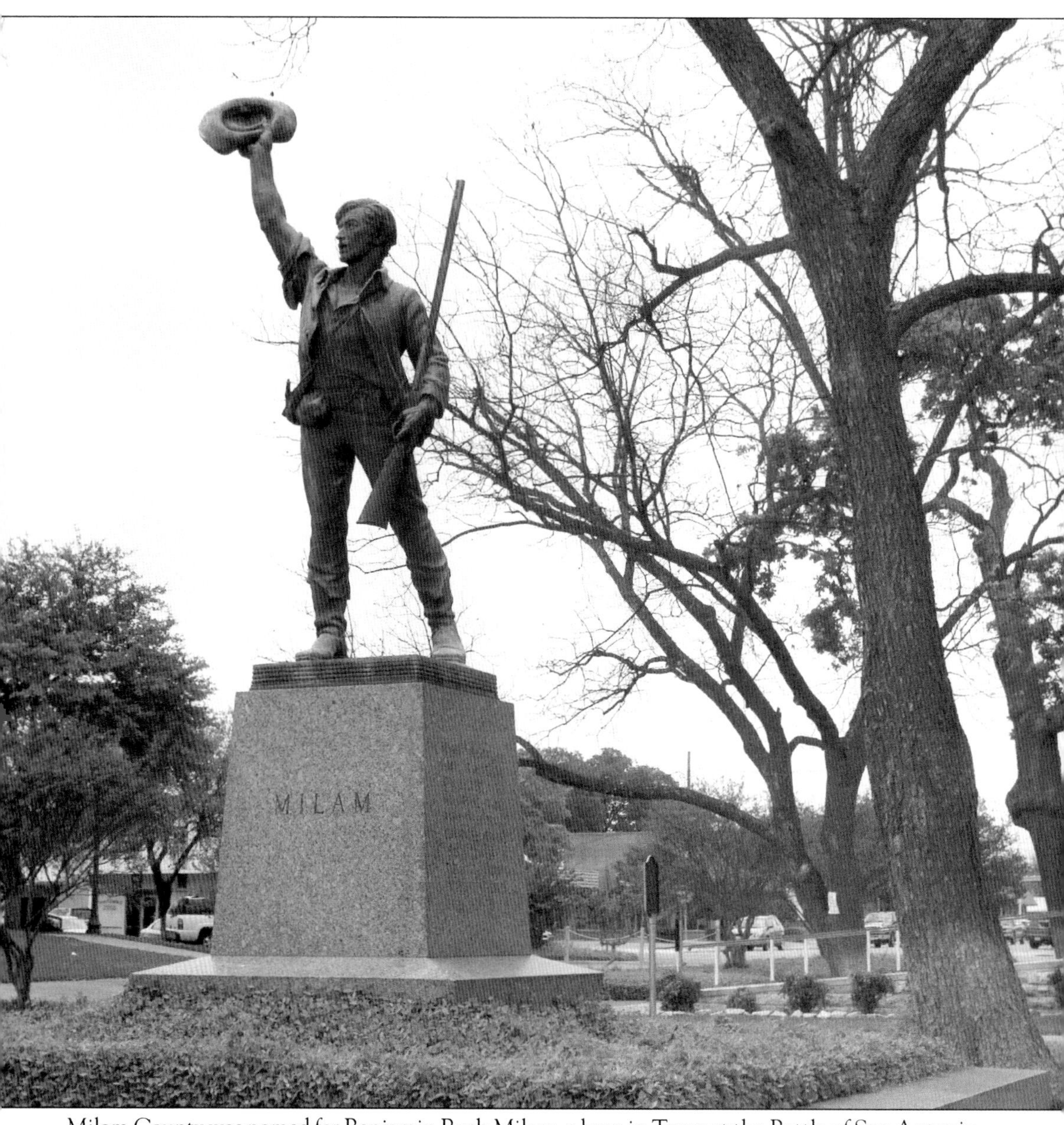

Milam County was named for Benjamin Rush Milam, a hero in Texas at the Battle of San Antonio in 1835. During this battle, he voiced his famous challenge, "Who will go with Old Ben Milam into San Antonio?" This challenge resulted in Milam's death, but ultimately in victory for the Texans. (Courtesy of Lucile Estell.)

One

Historic Bridges, Part One

At one time, Milam County had 27 large, historic bridges, the largest number of any county in Texas. They had emerged all about the countryside as the county came into full bloom in its present size and shape. Historic preservationists were among the population from the very first, but it was not until the mid-1990s that their attention turned to the historic bridges that crossed over rivers and streams throughout the area. Since that time, the bridges have become items of interest and study as well as preservation. It is difficult to know how to group the bridges for study because each one was built for different users with varying needs. As the authors began to divide the bridges, geographic location was one of the primary factors for inclusion and placement in each chapter. A prepared map has been added that will be helpful to visitors to Milam County who have a yen to visit and view one or more of these old treasures.

Included in chapter 1 are the following bridges: Worley Bridge, McClaren/Marlow Bridge, Bryant Station Bridge, Brushy Creek/Salty Brushy Creek, and Faubion/Sugarloaf Bridge. The Texas Department of Transportation worked with the Milam County Commissioners Court to fully restore Worley Bridge, and it is in use today. One may begin a tour of bridges at any location, but one reason for beginning with Worley Bridge is that it is located at historic Apache Pass, an important visitor site along El Camino Real de los Tejas National Historic Trail. It is the newest National Historic Trail in the United States (Historic Trail No. 19).

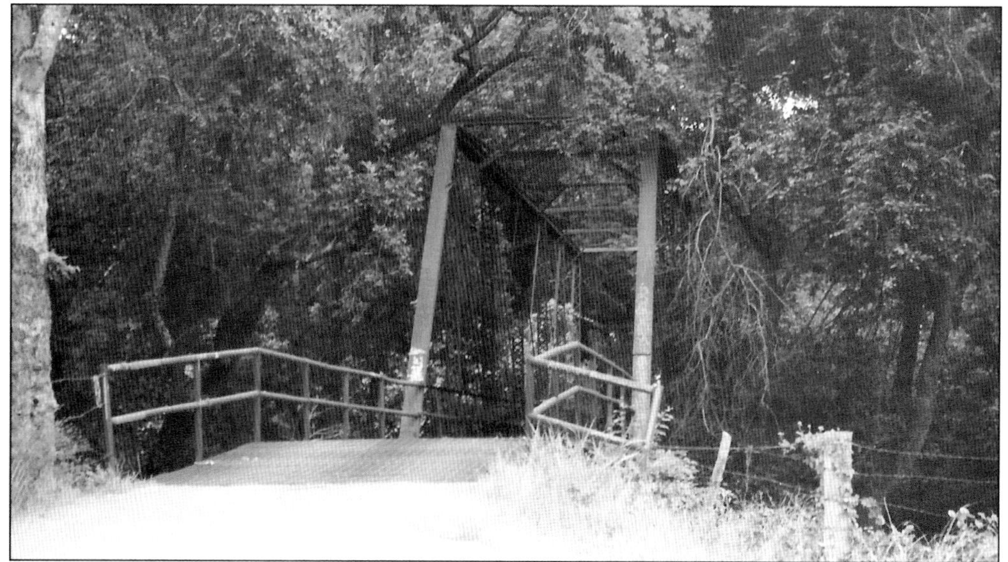

Worley Bridge, a grand old structure, still majestically spans the San Gabriel River near Apache Pass. This Pratt through pinned bridge, begun in 1911, is 271 feet long, spans 113 feet, and has an overall width of 13 feet, with steel grate decking. This is the only one of the historic bridges in Milam County that is still used as a traffic bridge, for automobiles and trucks of limited weight. (Courtesy of Joy Graham.)

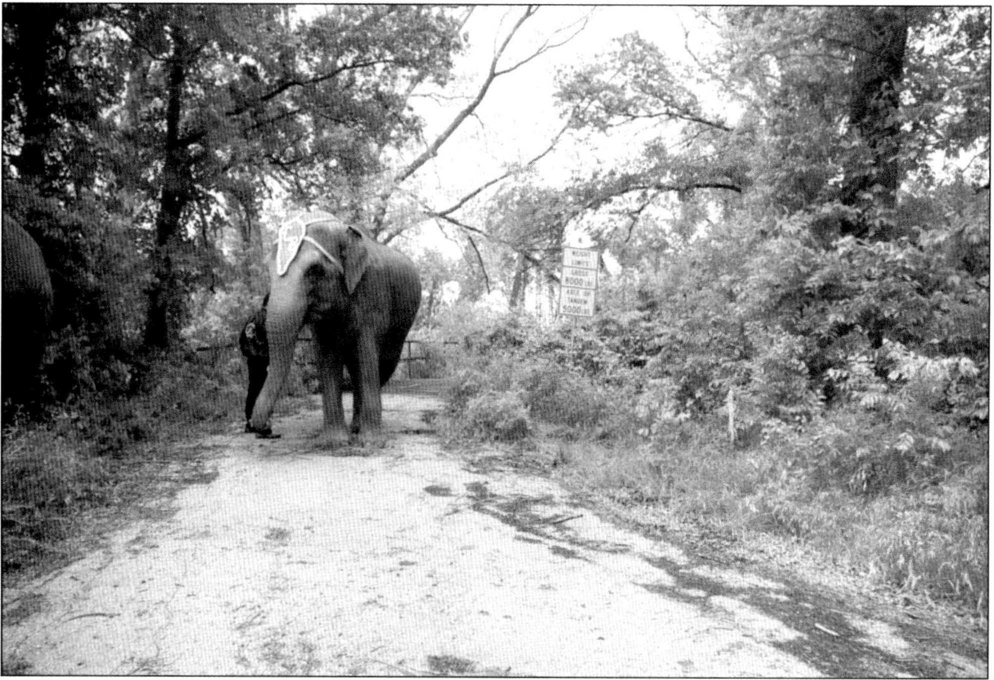

Elephants roam at will in the historic Apache Pass crossing area on a delicious summer day in 2013. A circus that was traveling through the county released these handsome creatures for a brief romp in the park. Their majestic size matched the majesty of Worley Bridge. The area where they visited has been inhabited by Native Americans as well as Spanish and French explorers. (Courtesy of Lucile Estell.)

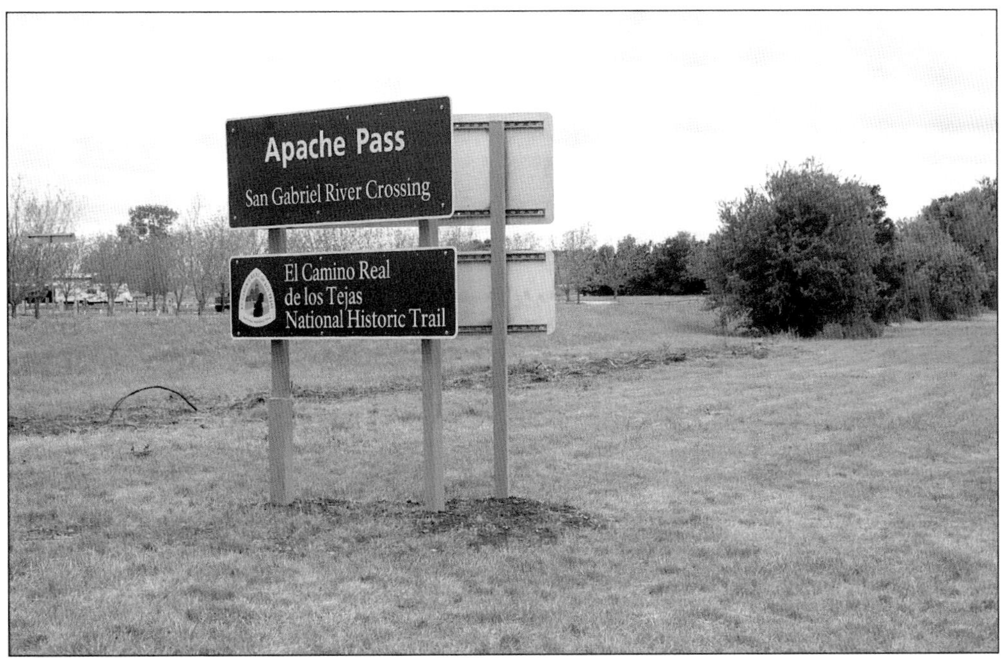

In 2004, Congress recognized a new historic trail, El Camino Real de los Tejas National Historic Trail. An important segment of this trail is located in the vicinity of this historic bridge. This was the site of three missions and much activity during the 18th century. The trail is marked throughout this area of Texas. (Courtesy of Lucile Estell.)

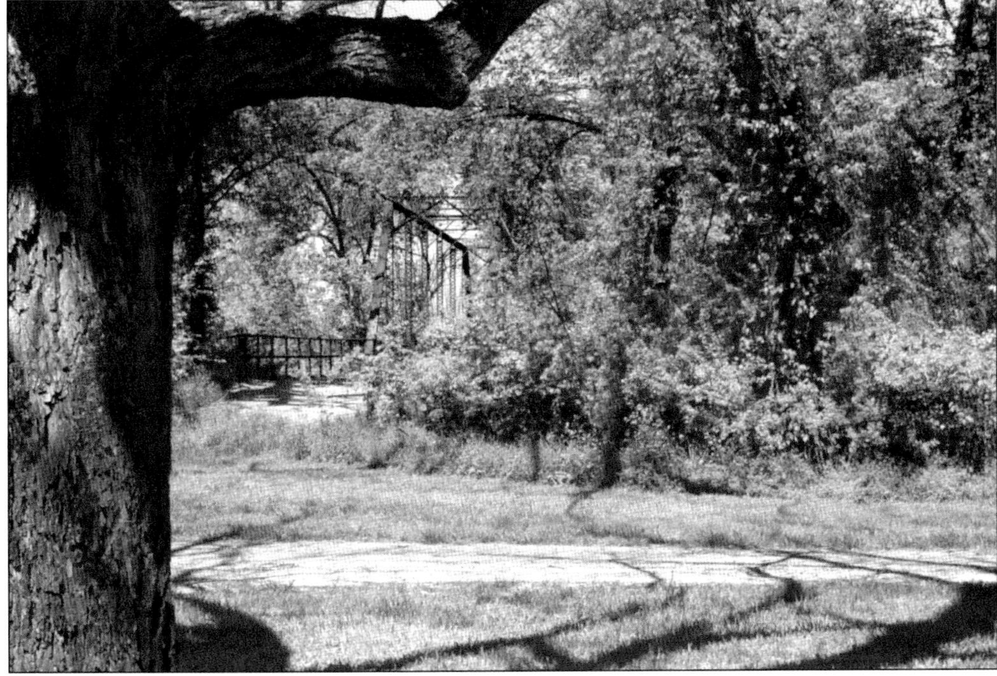

Worley Bridge is seen here through the greenery at Apache Pass. Traffic uses this scenic route daily as visitors and locals enjoy the beauty of the area as well as the site's historical significance. (Courtesy of Lucile Estell.)

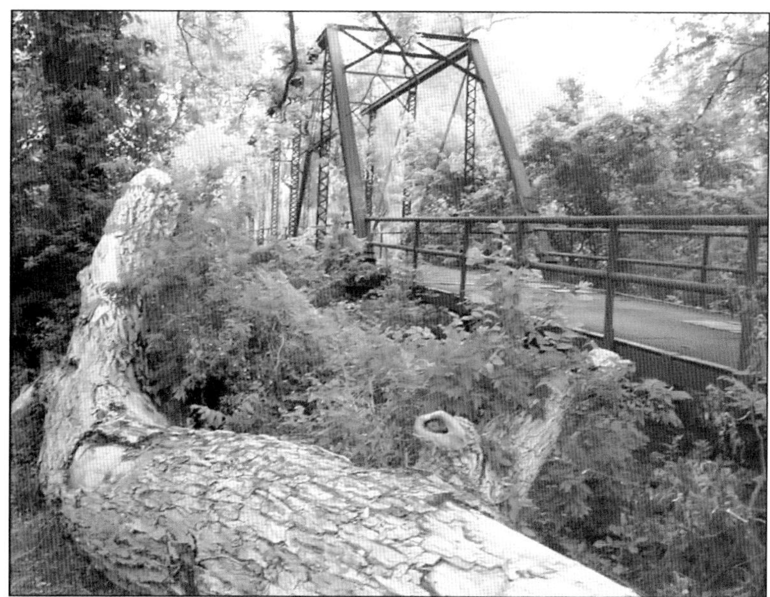

This photograph of Worley Bridge was taken after a thunderstorm had passed through the area. Note the thick foliage and the fallen tree. Worley Bridge has been a spectator to all kinds of weather at Apache Pass. (Photograph by Mike Brown, courtesy of *Rockdale Reporter*.)

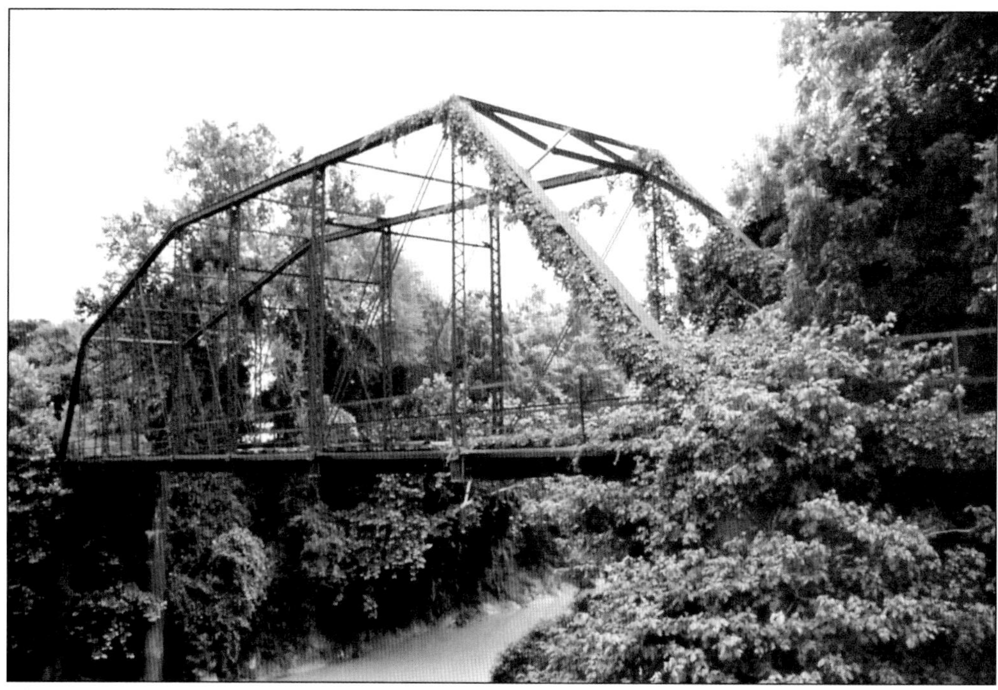

The commissioners court ordered the McClaren Bridge to be built at a cost of $5,000. Several people in the community contributed cash toward the building of the bridge. Included among the donors were the following people: S.W. McClaren Sr. gave $175, Giles McDermott gave $150, John Hobson contributed $20, Ernest Watson donated $10, Brice Burnett gave $5, and Ed Tindall donated $2.50. (Courtesy of Joy Graham.)

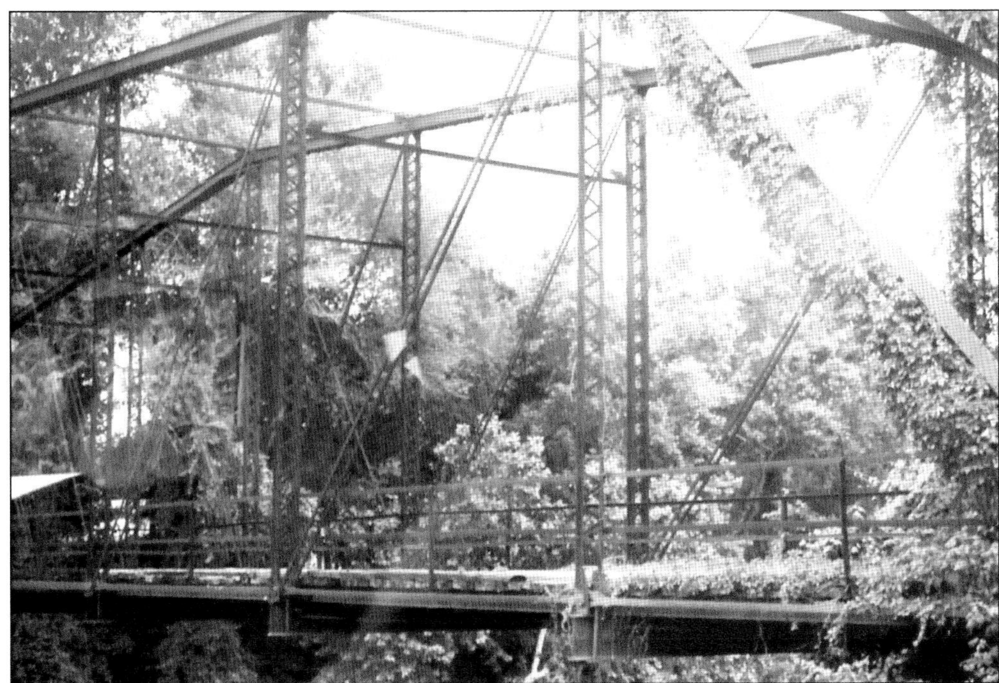

In October 1911, it became apparent that a bridge was needed over Little River in the Marlow community to provide a crossing of the river on the Nashville Road to Cameron, which was an important traffic and trade route during the 19th century. Before the bridge was built, a public ferry operated at this crossing. This bridge is often called the Marlow Bridge because it is located in the community of Marlow. (Courtesy of Joy Graham.)

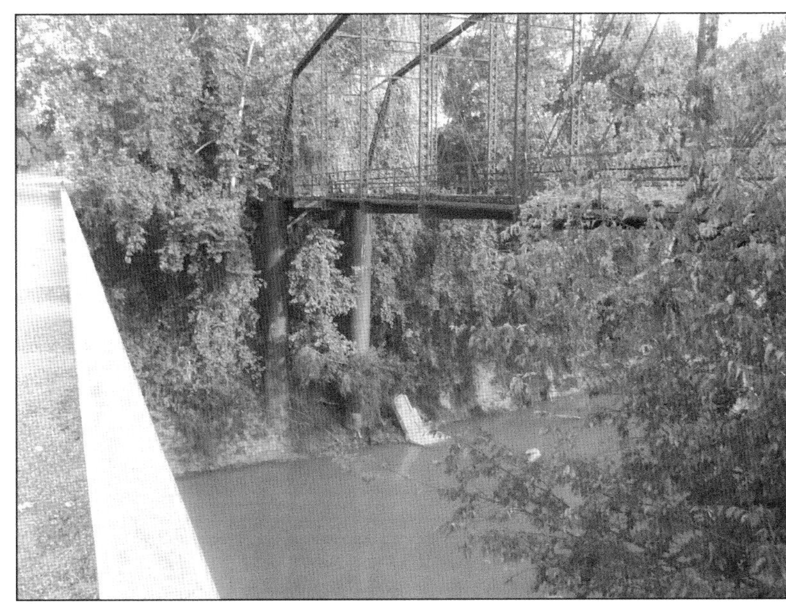

During the late 1980s, the McClaren Bridge became unsafe for vehicular traffic. It was replaced with a cement slab. (Courtesy of David Galbreath.)

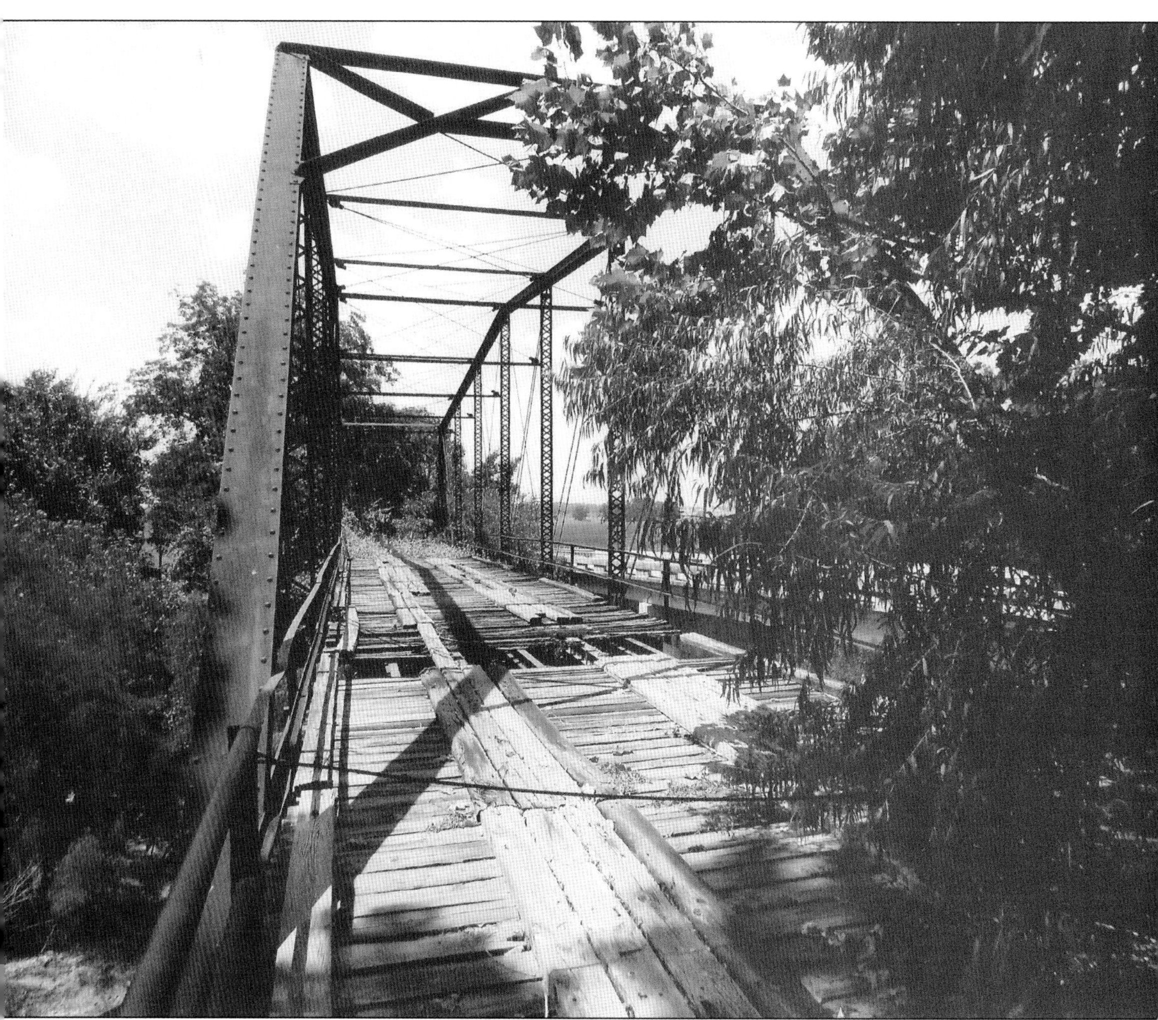

This bridge was needed in order to help Marlow residents get their crops to market using a more direct route. The place chosen for the bridge was property known as Lamkin's Crossing because it was a part of the extensive acreage owned by the W.S. Lamkin family. The Marlow community continues to flourish, and the original bridge stands as a reminder of days past. Today, the road past it continues to be a busy thoroughfare. (Courtesy of Dolores Mode.)

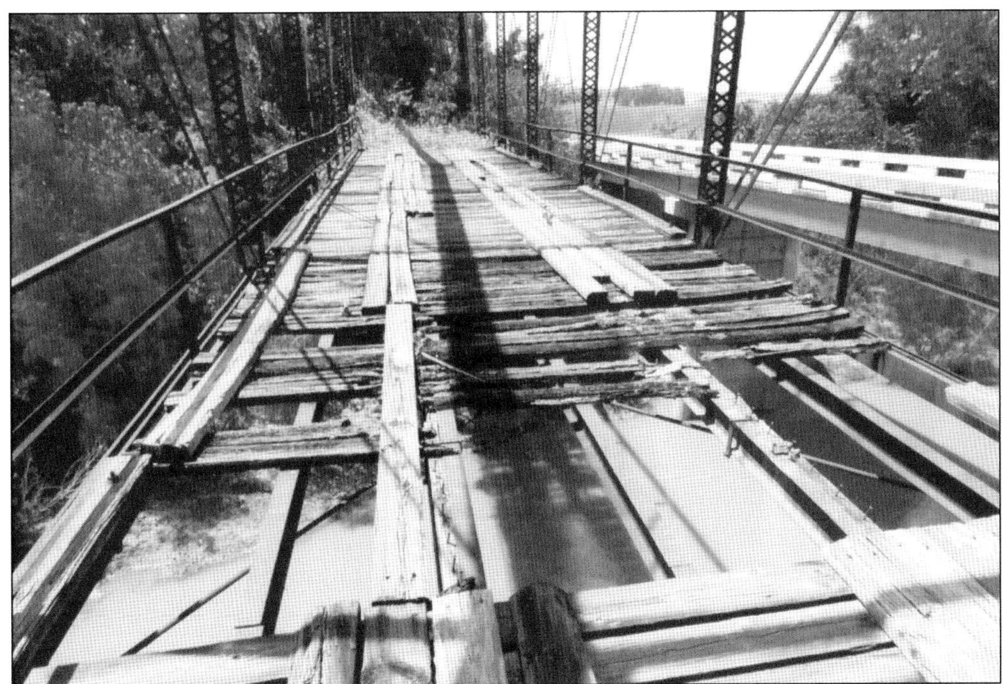

W.B. Horton was general contractor for building this bridge, with S.W. McClaren as supervisor. Called both McClaren Bridge and Marlow Bridge, it is no longer used for vehicular traffic. (Courtesy of David Galbreath.)

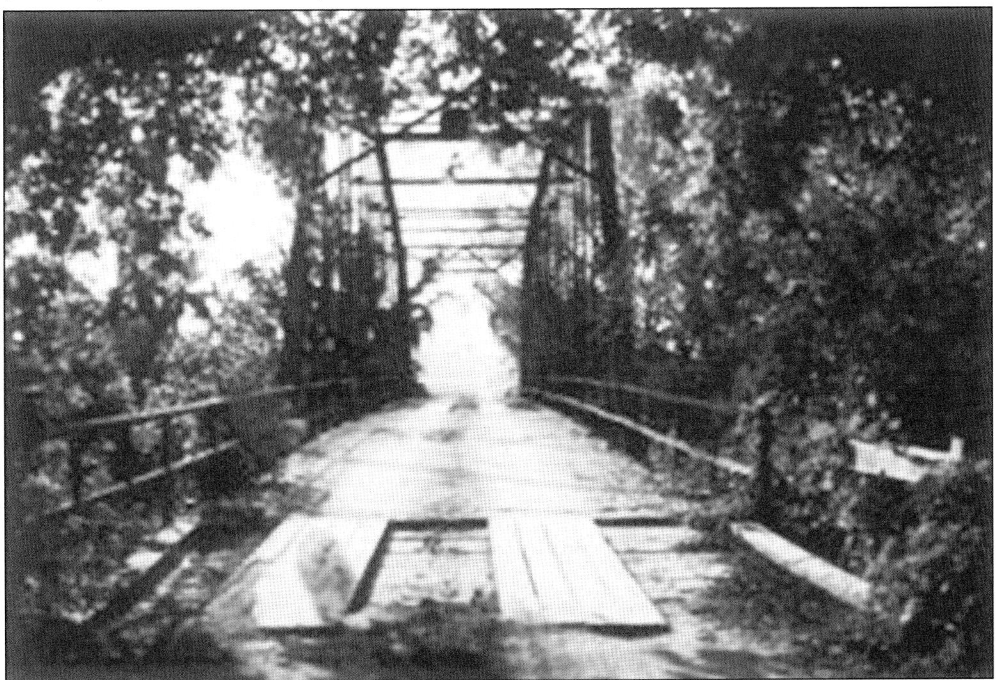

Bryant Station Bridge is located on County Road 106 near Buckholts, which is in the northwestern part of the county. It was an important road during the time when people traveled from Marlin to Austin by stagecoach. (Courtesy of the Milam County Historical Museum.)

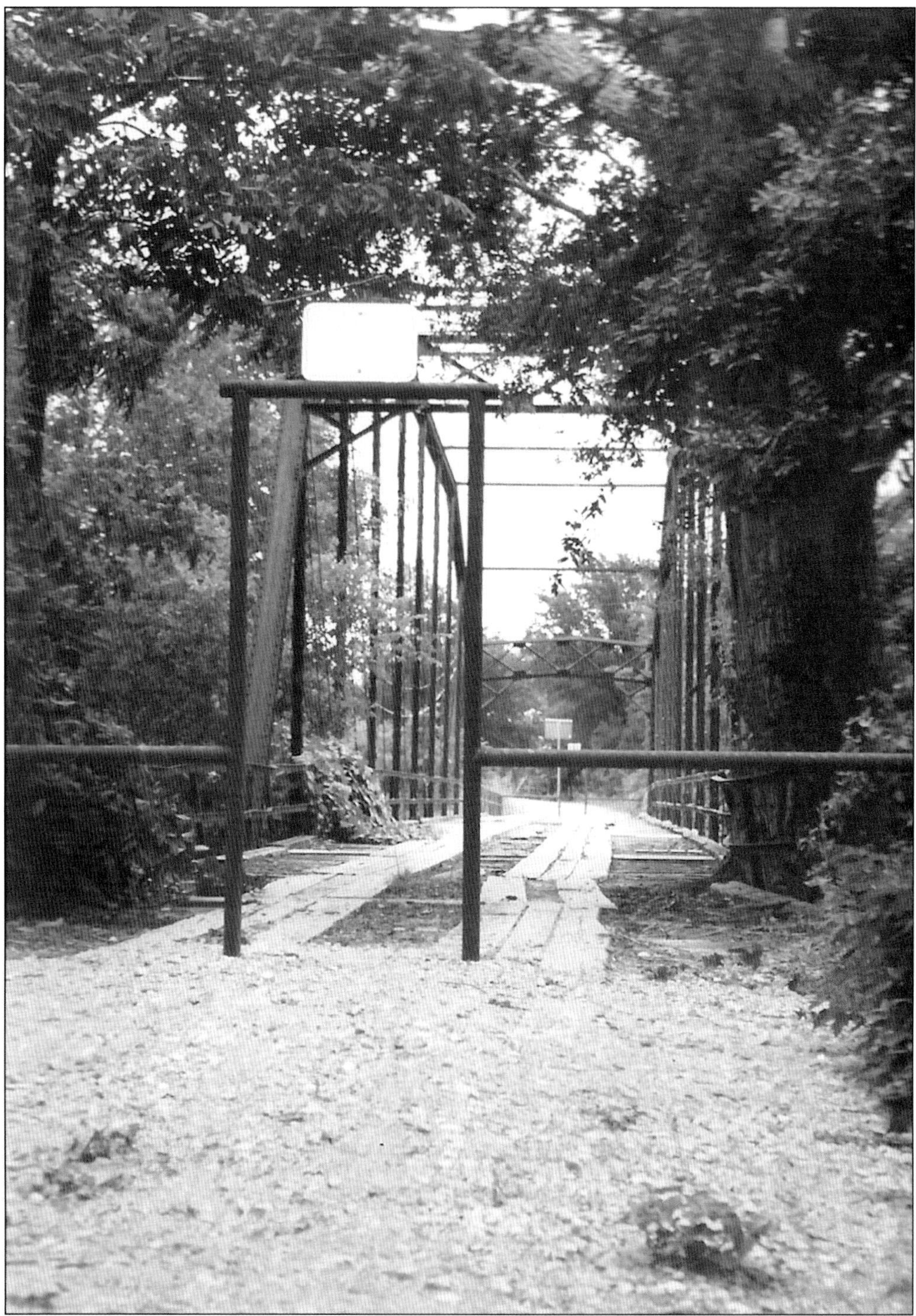

Bryant Station was named after the US Army station that was established by order of Sam Houston, who was the president of the Republic of Texas. The station was built by Benjamin Bryant for the purpose of keeping Native Americans out of the area and away from the then capital at Washington-on-the-Brazos. (Courtesy of Lucile Estell.)

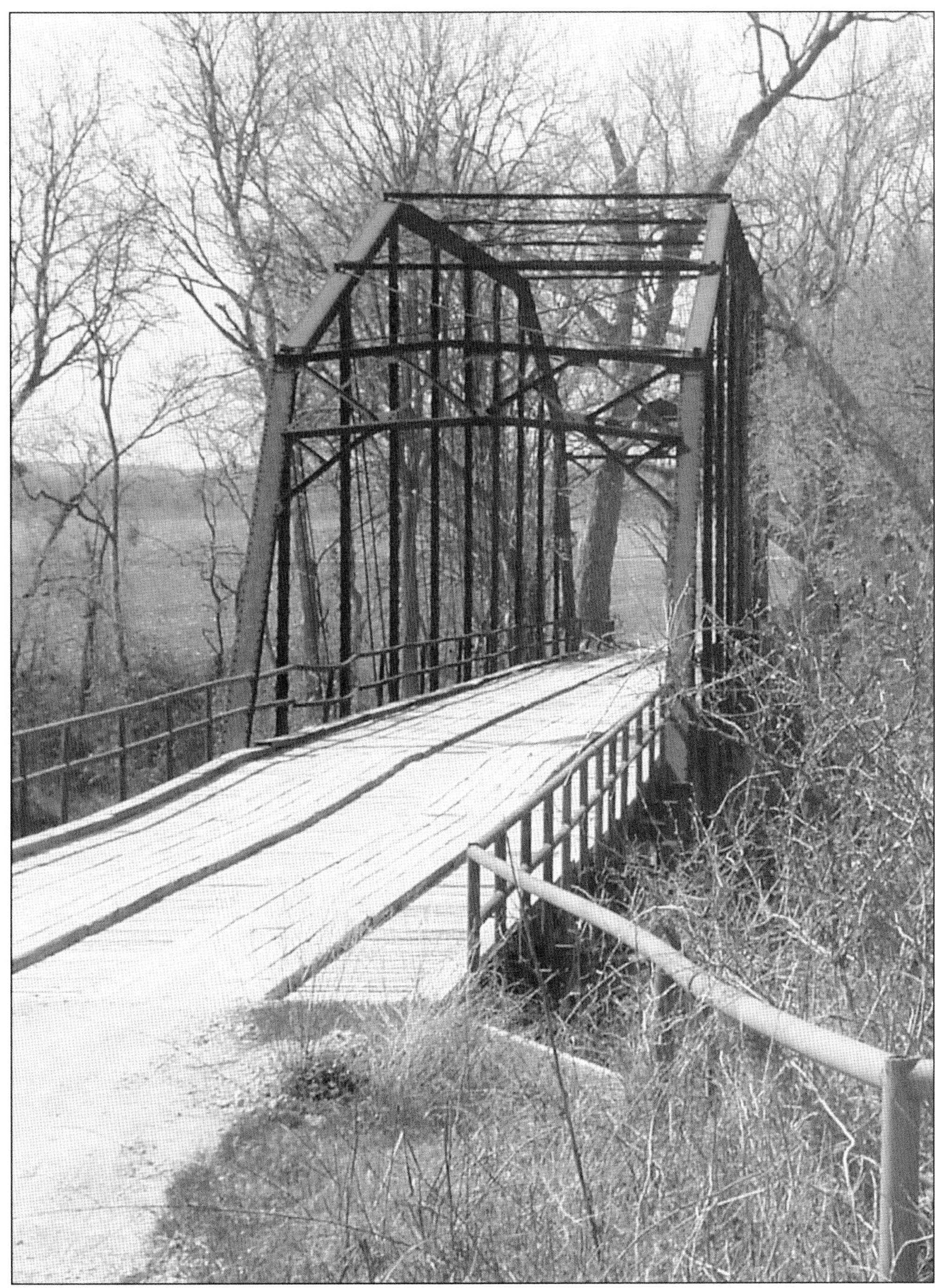

The bridge was built in 1909 as a camelback truss bridge by C.Q. Horton of Austin and the Chicago Bridge and Iron Company for $5,980. This bridge is no longer in use, but has been replaced by a 420-foot-long concrete structure that is 32 feet wide. The original bridge, one of two existing of this type of camelback bridge, is now used as a pedestrian bridge. (Courtesy of David Galbreath.)

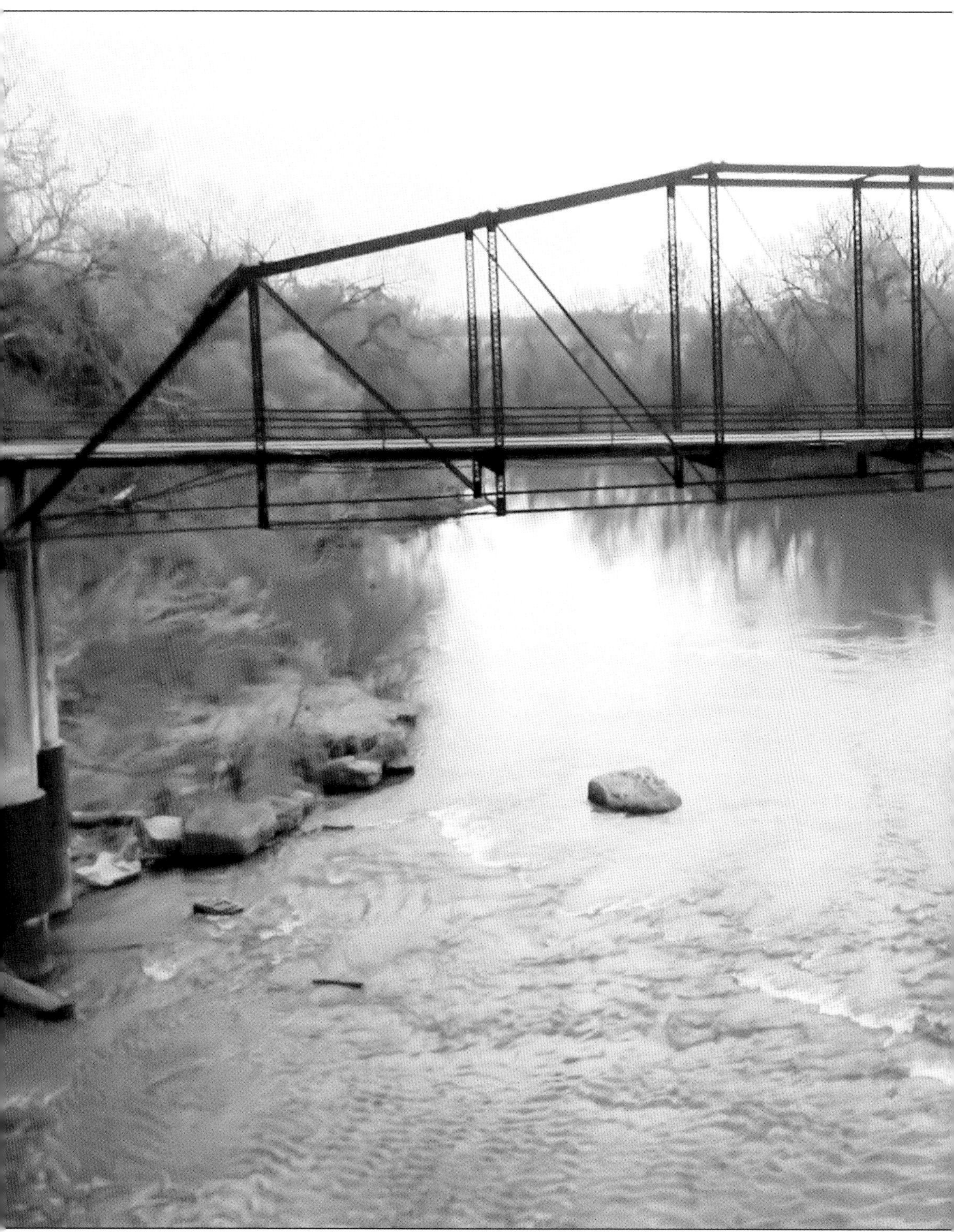

This is a beautiful scenic view of Bryant Station Bridge spanning the Little River. It crosses the river just south of Bryant Station. Its primary purpose was to connect Buckholts, Davilla, and

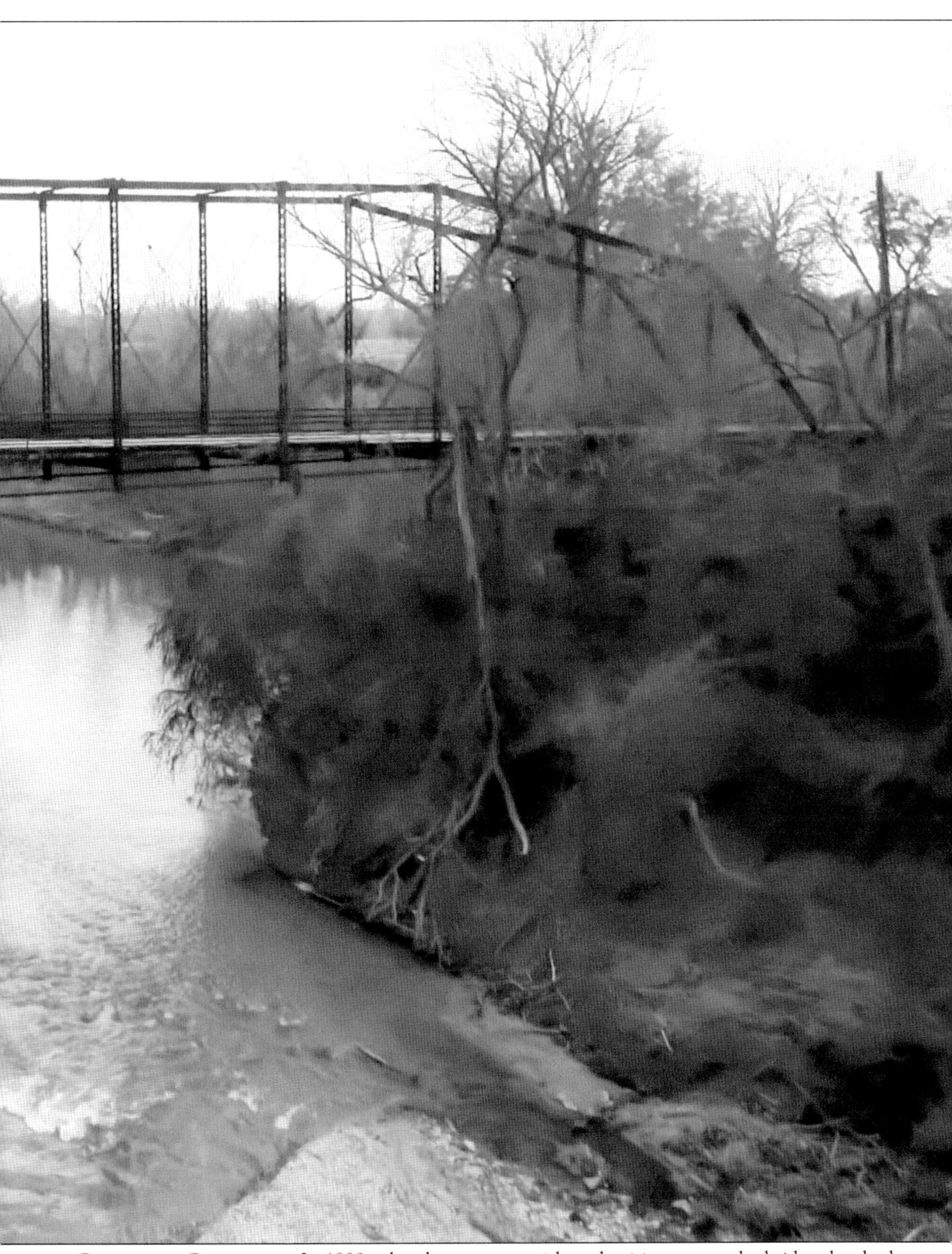

Cameron to Georgetown. In 1999, a local group met with authorities to save the bridge that had been spanning the Little River for 90 years. (Courtesy of David Galbreath.)

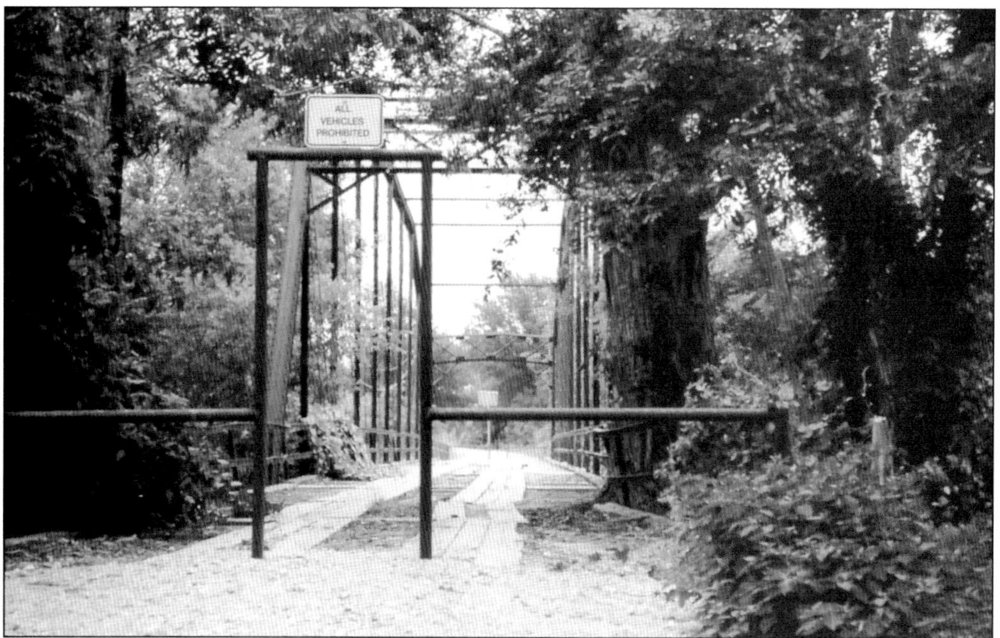

The bridge was not demolished, but refashioned for pedestrian traffic. A few large trees had to be cut down, but nothing else of value had to be destroyed. (Courtesy of Joy Graham.)

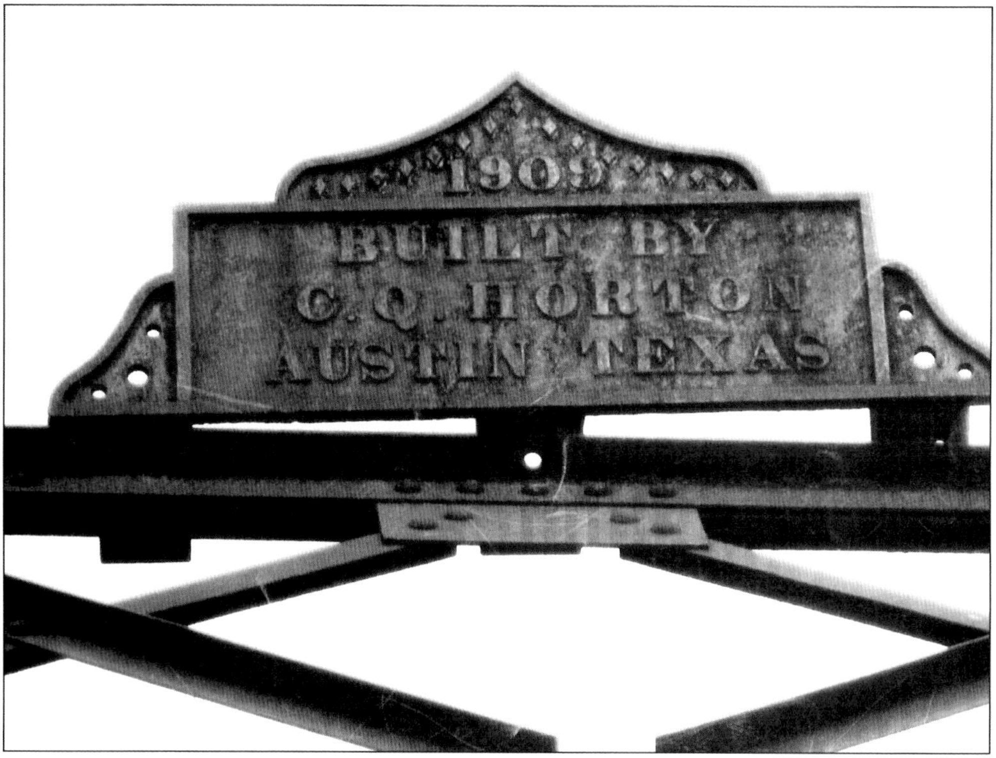

This photograph shows the Bryant Station Bridge plaque dated 1909. It was built by C.Q. Horton of Austin, Texas. This company was responsible for much bridge work in Milam County. The new bridge goes just west of the old span. (Courtesy of Jack and Beth Brooks.)

This photograph shows yet another view of the Bryant Station Bridge over the Little River. The bridge is no longer being maintained by the county. It was, after its closure to vehicular traffic, reconfigured as a pedestrian bridge. (Courtesy of the Milam County Historical Museum.)

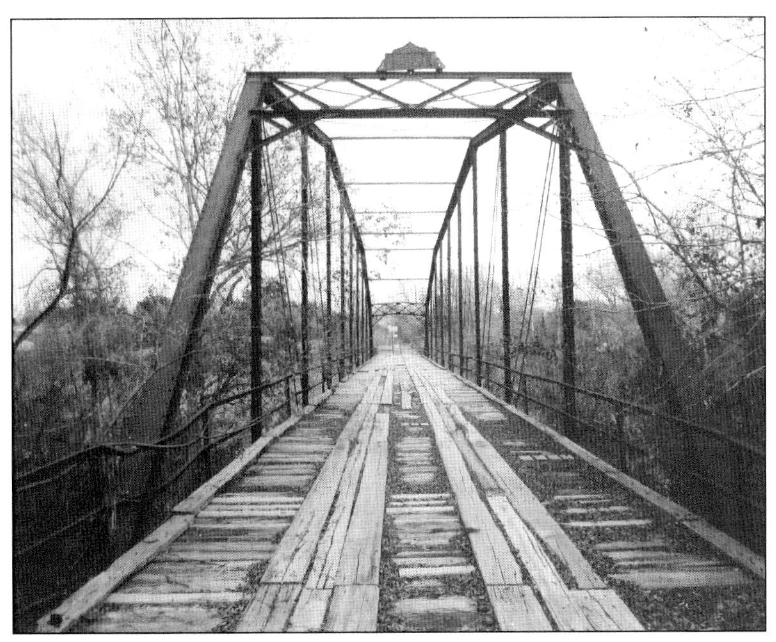

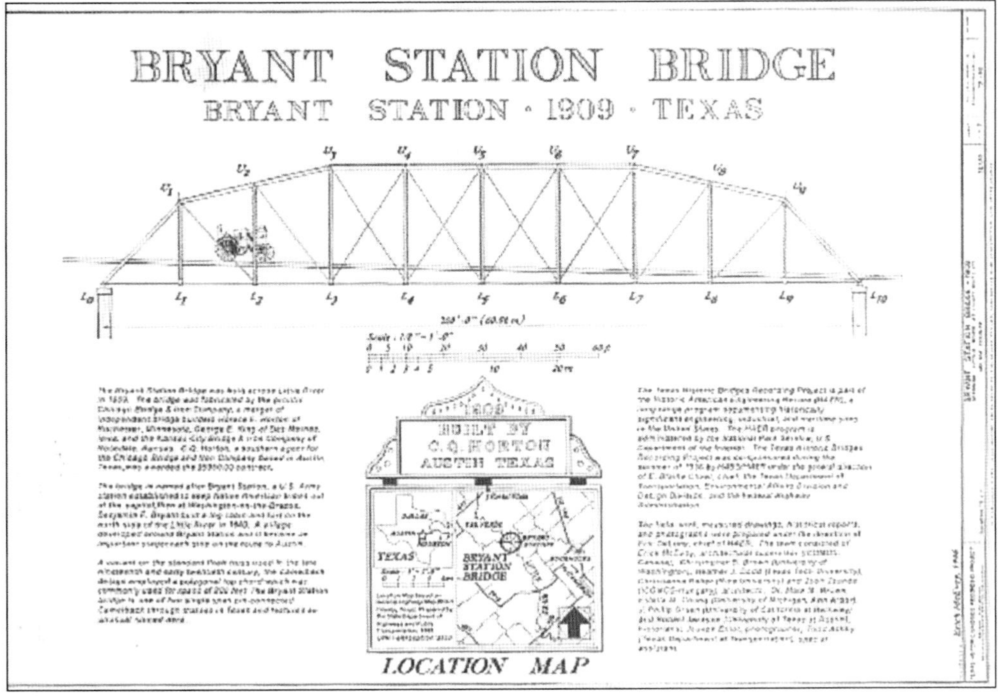

This photograph shows the specifications for the original Bryant Station Bridge that was built in 1909. Its structural integrity lasted for decades. This drawing is from the Historic American Engineering Record. (Courtesy of Texas Department of Transportation.)

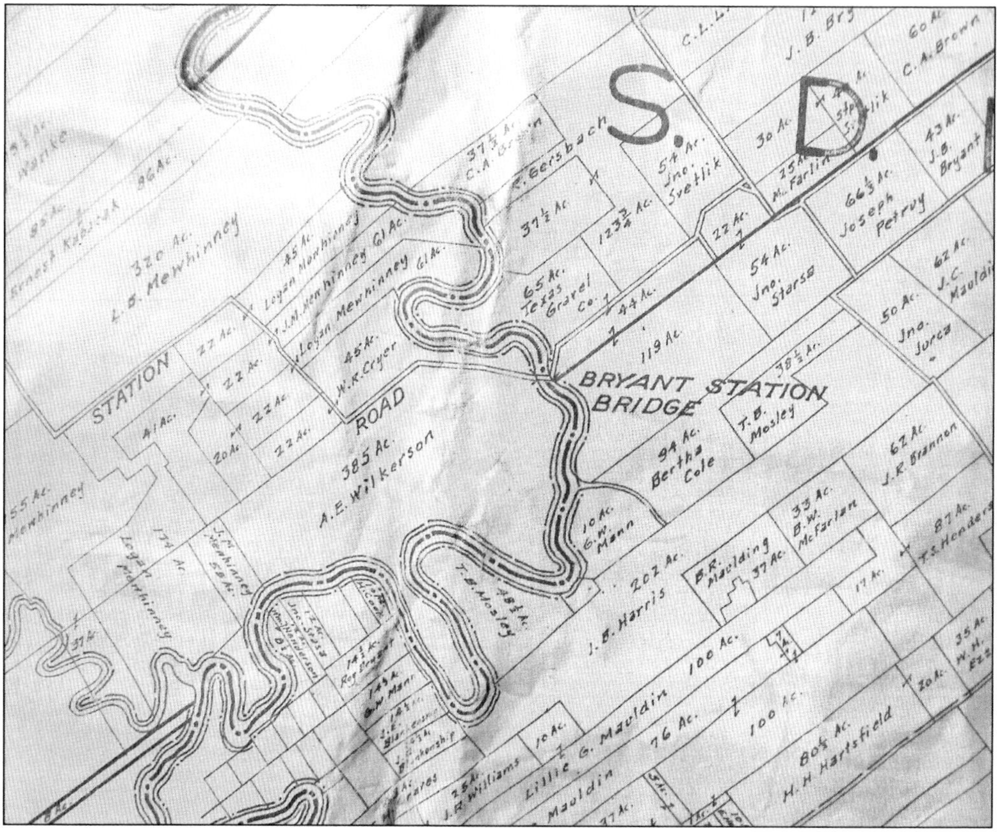

This is a copy of a Milam County map showing the exact location of the Bryant Station Bridge and the surrounding area. The map emphasizes the fact that the Little River meanders considerably as it moves through the landscape of the county. (Courtesy of Jack Brooks.)

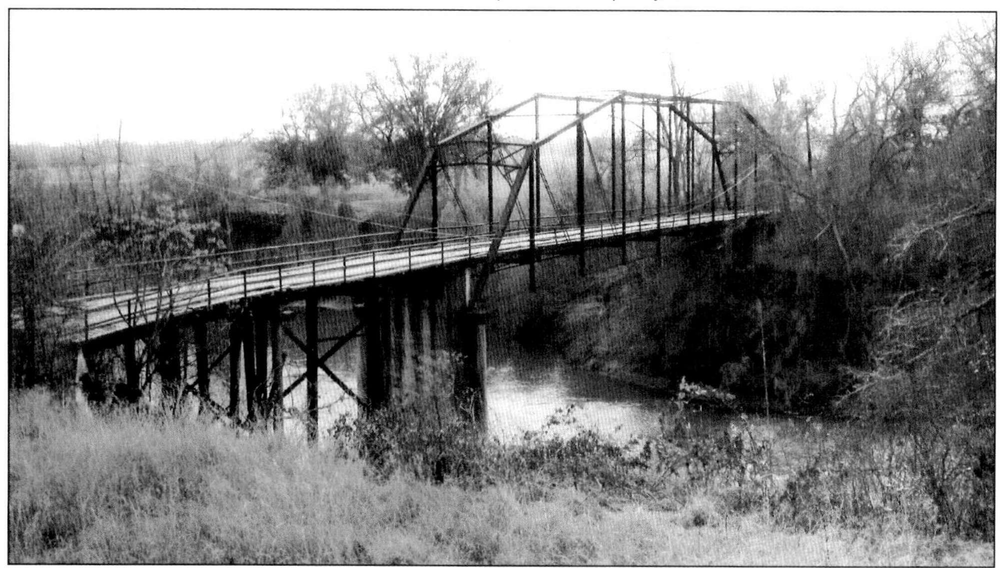

The Bryant Station Bridge was as unique as it was historic. It provided safe crossing to many generations of county residents. (Courtesy of David Galbreath.)

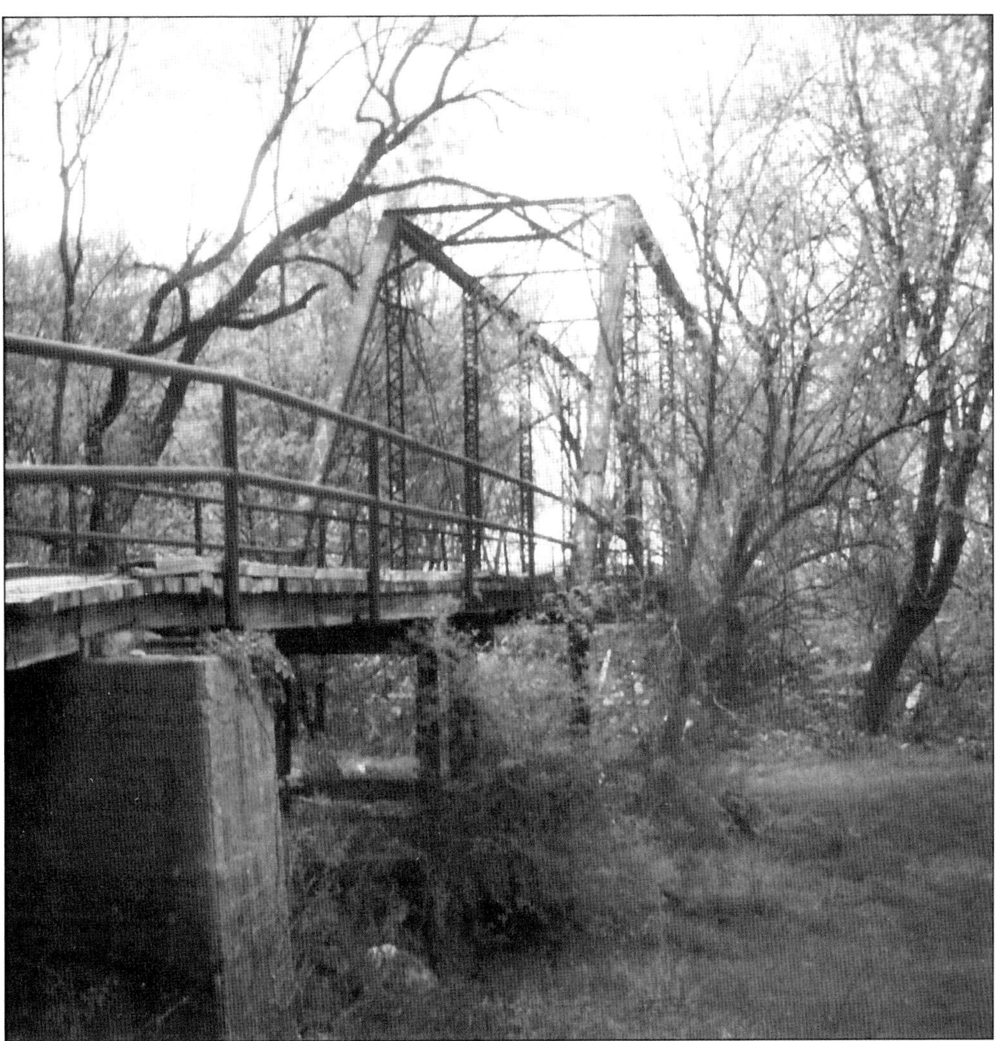

Here is a view of the entrance to Brushy Creek/Salty Brushy Creek Bridge. In 1870, the community of Salty was established. The settlement got its name from the Salty Creek flowing nearby. In 1910, the Milam County Commissioners Court voted to build a bridge on County Road 440 over Brushy Creek, 4.5 miles southeast of County Road 486 south of Thorndale, Texas. This bridge became known by three names: Brushy Creek Bridge, Salty Brushy Creek Bridge, and Ledbetter Bridge. In March 1999, commissioners voted to replace the Brushy Creek Bridge with a concrete slab bridge. (Photograph by Jack Tumlinson, courtesy of Tense Tumlinson.)

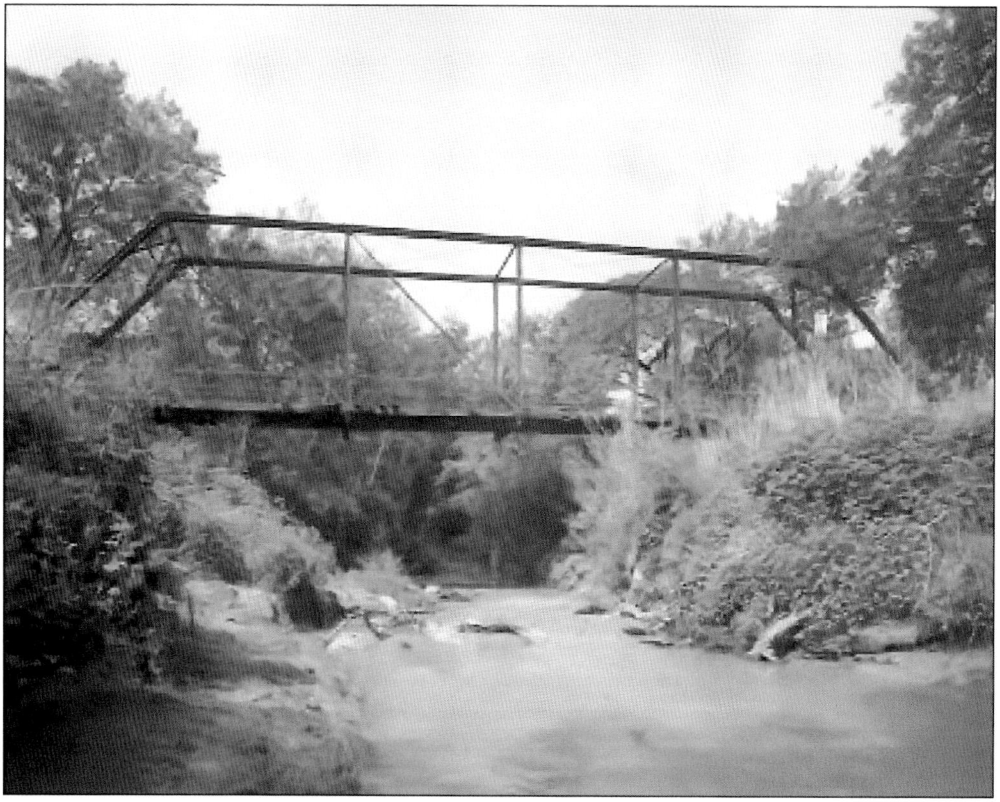

This photograph offers a side view of Brushy Creek/Salty Brushy Bridge when it was located on County Road 440. In November 1910, a contract to build the bridge was issued to C.Q. Horton Bridge Co. of Austin, Texas. The construction was to be a Pratt through pinned truss bridge that was 133 feet of total length, with a 13.8-foot-wide deck. (Courtesy of Library of Congress.)

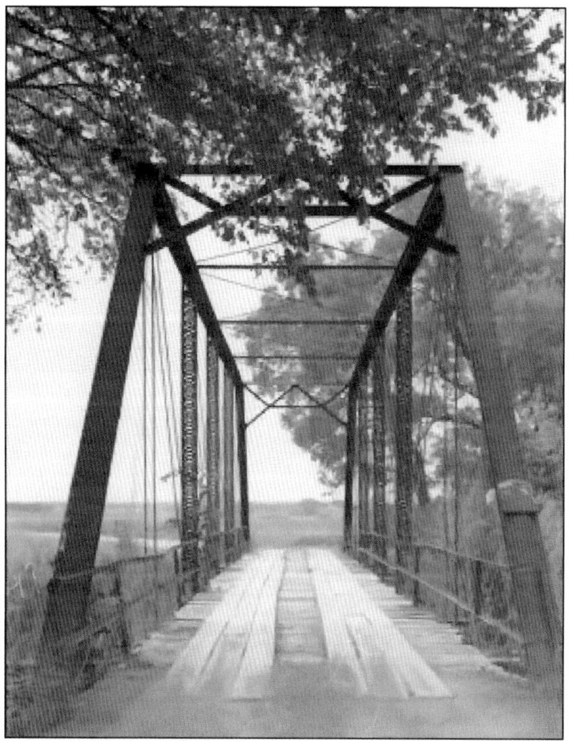

In the early 1900s, when Brushy Creek Bridge was built, roads were becoming more important to the farm economy. Farmers in Milam County used farm machinery, produced crops above subsistence level, and needed better roads to transport their goods to market. Located on a local road just south of Thorndale, the Brushy Creek Bridge connected farmers with the Missouri Pacific Railroad depot there. (Courtesy of Library of Congress.)

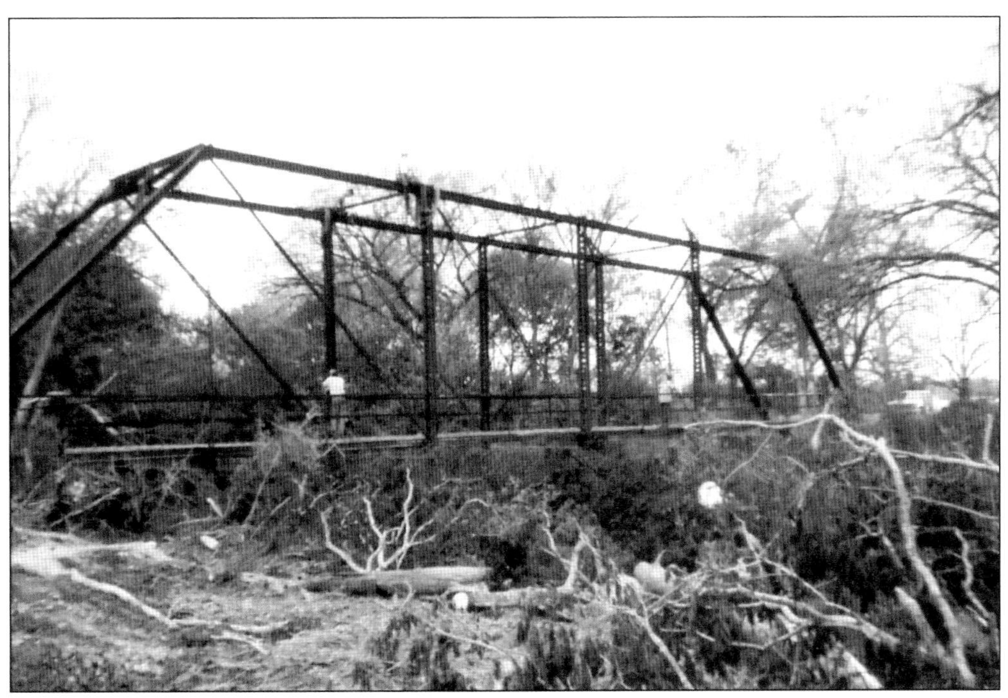

Above, the Davenport Mammoet workers of Rosharon, Texas, are cutting down the trees beside the historic bridge and preparing to lift the Brushy Creek Bridge off its old foundation. The photograph below shows the bridge being removed. The bridge enabled travel in Milam County for 88 years. (Both, courtesy of Dolores Mode.)

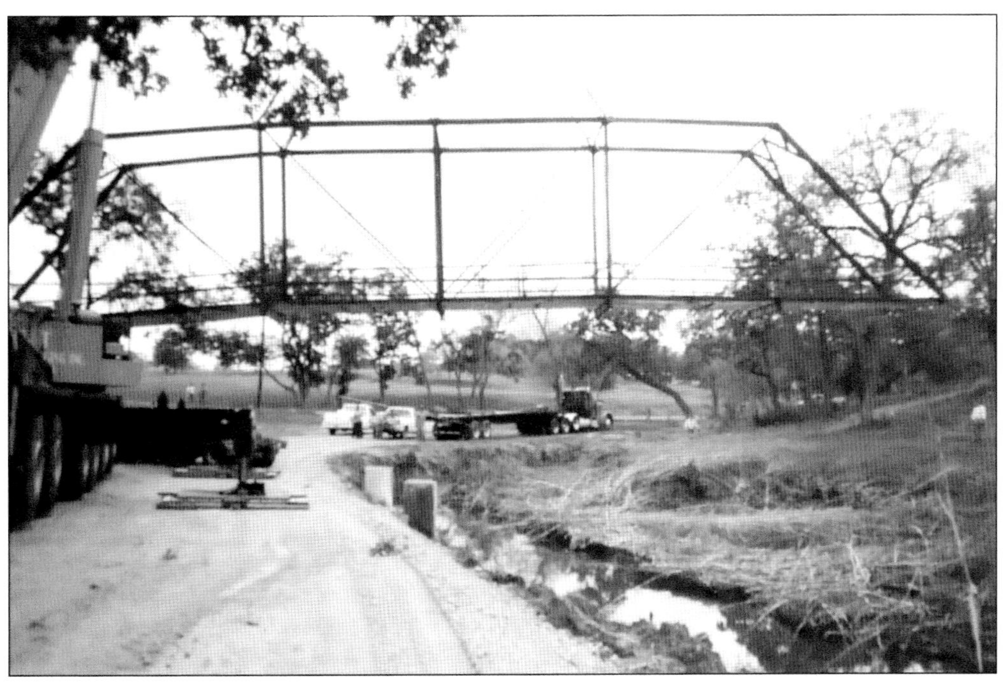

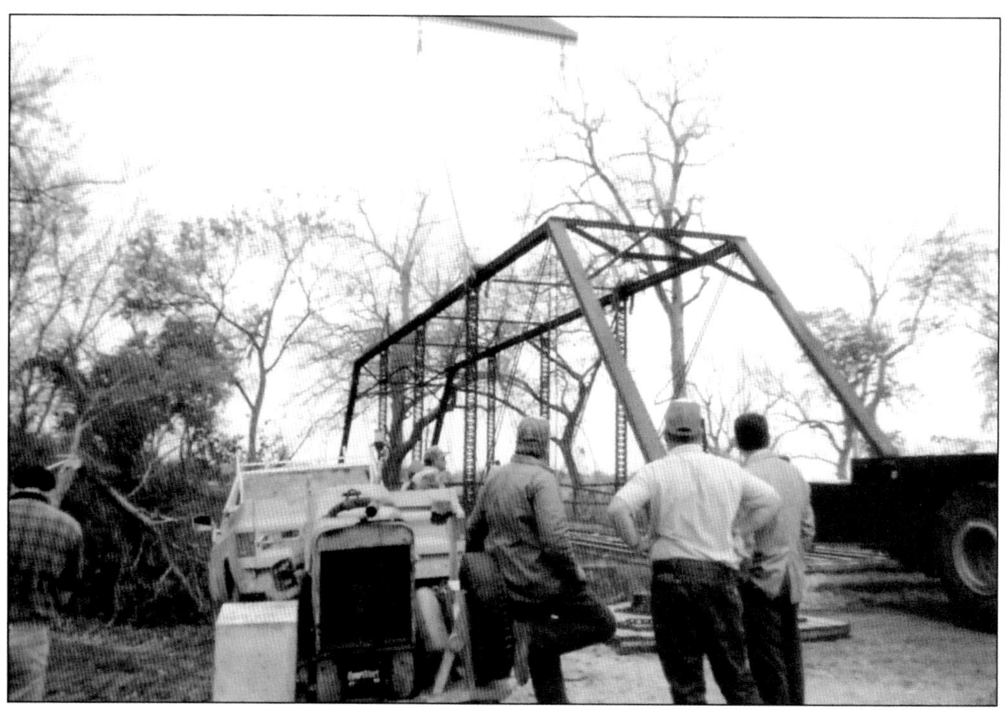

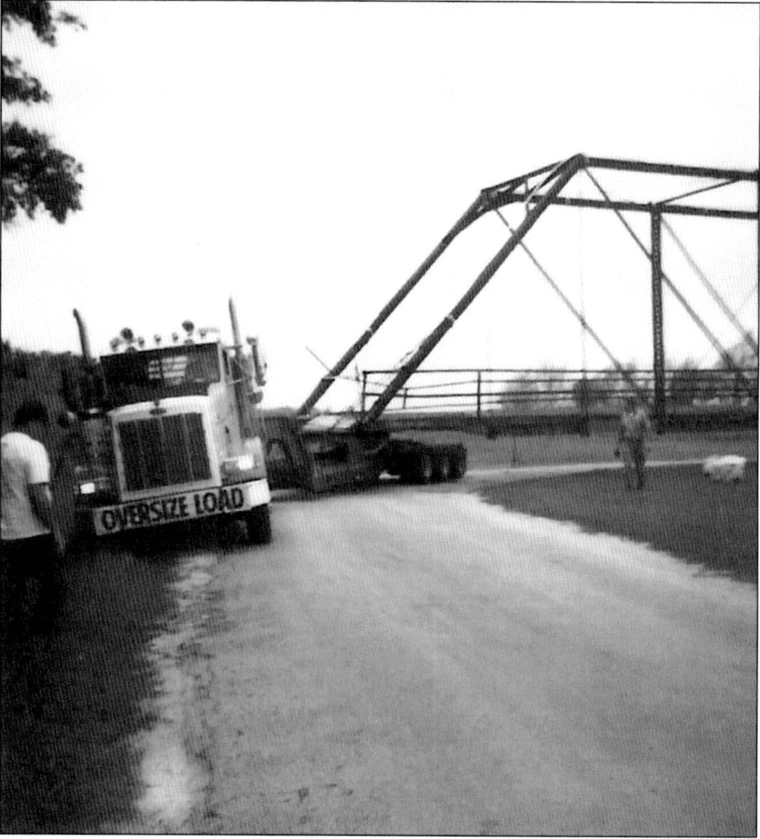

The photograph above shows workers and sightseers out watching a crane lift the historic Brushy Creek Bridge. The image at left was taken on March 23, 1998. Employees of Davenport Mammoet of Rosharon, Texas, worked through the day moving the Brushy Creek/Salty Brushy Bridge. The bridge is rounding the corner onto County Road 440 Thorndale, Texas. (Above, courtesy of Dolores Mode; left, courtesy of Joy Graham.)

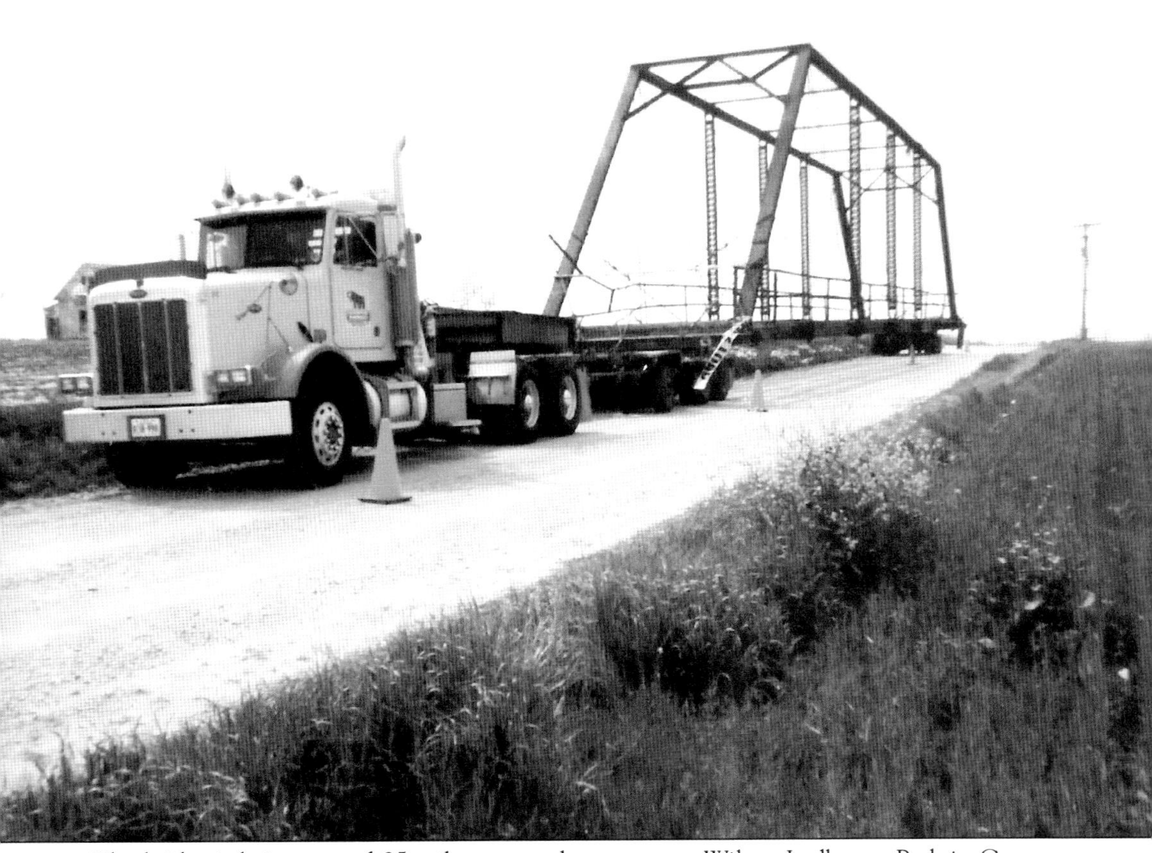

The bridge is being moved 25 miles across the county to Wilson-Ledbetter Park in Cameron, Texas. Brushy Creek Bridge is on its way to its new home. (Courtesy of Dolores Mode.)

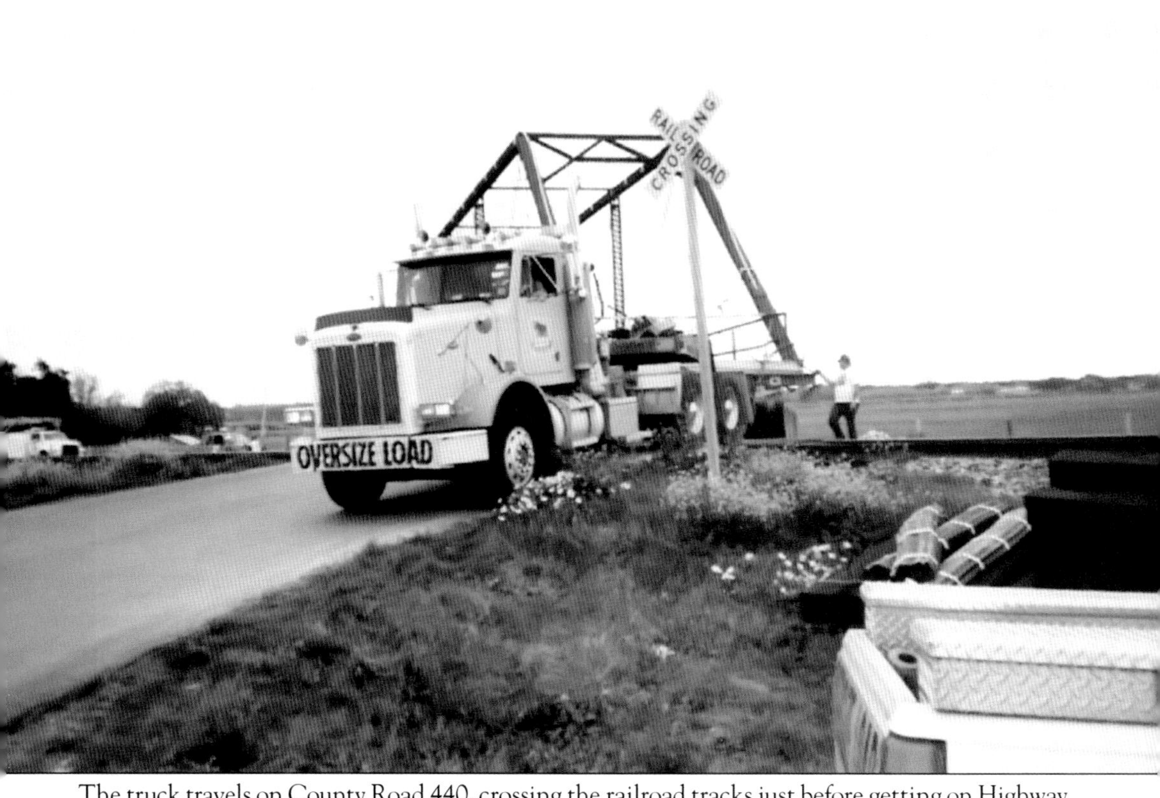

The truck travels on County Road 440, crossing the railroad tracks just before getting on Highway 79. In the photograph, many trucks follow this oversize load. (Courtesy of Dolores Mode.)

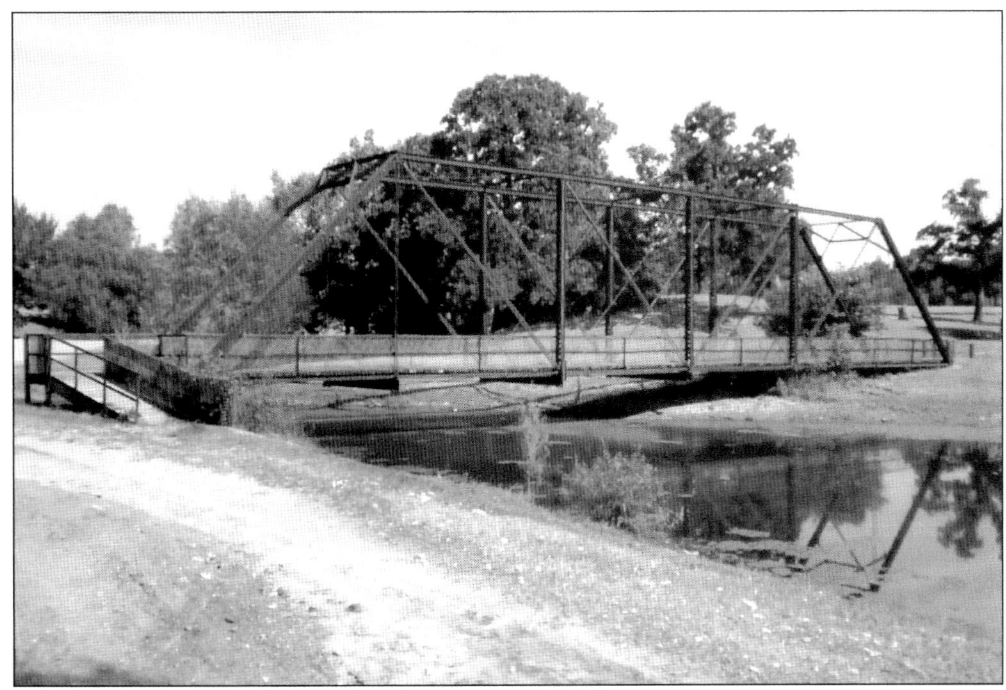

After a week of delays due to utility clearance, Brushy Creek/Salty Brushy Bridge finally made its way to its new home. On April 1, 1998, it was placed across a section of the lake in Wilson-Ledbetter Park. The bridge was restored as a pedestrian bridge. Two entrances were modified and are used for foot traffic and social events. Now, many of the local residents refer to the bridge as Ledbetter Bridge. In September 2016, Wilson-Ledbetter Parks celebrated its centennial anniversary. (Courtesy of Joy Graham.)

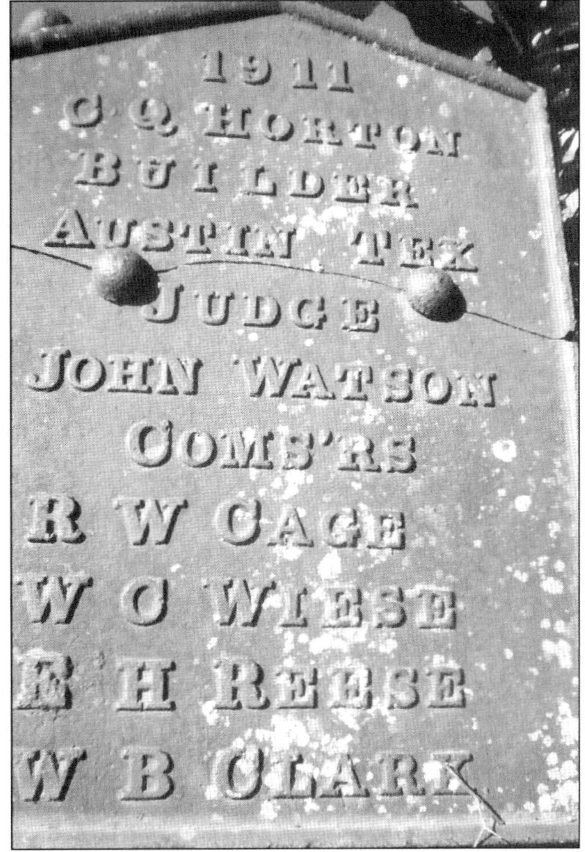

The bridge proudly displays a plaque dated 1911 indicating the completion date. There is a crack in the plaque. The historic bridge has endured years of exposure to the outside elements. (Courtesy of Joy Graham.)

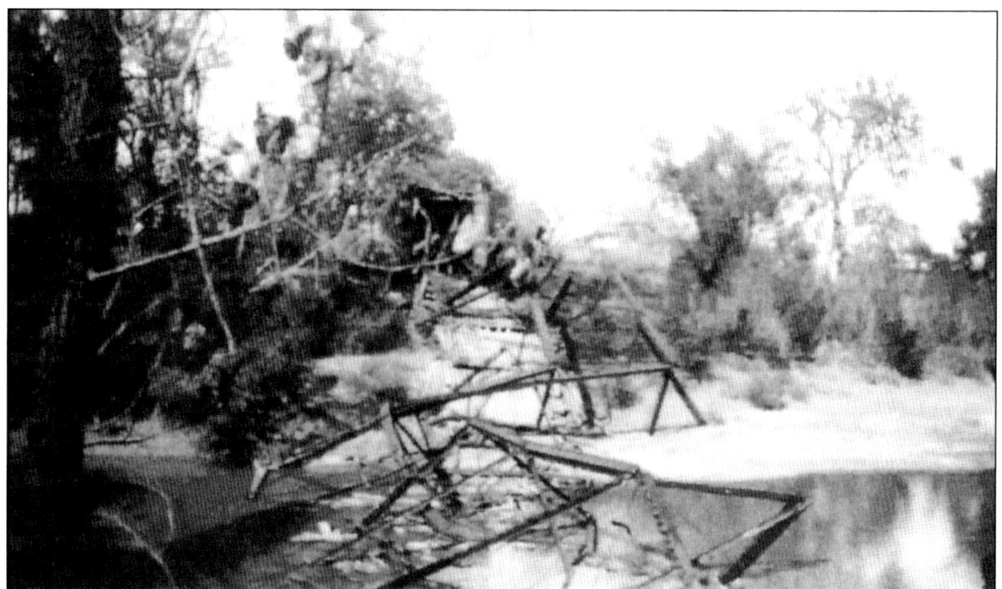

The Faubion Bridge was built in 1890. By 1902, the George E. King Bridge Company of Des Moines, Iowa, was contracted to repair the structure. In 1905, the bridge fell into the river again. In 1906, E.P. Alsbury and Son of Houston was contracted by the county to erect a new bridge at the same site. Due to erosion of the center cement pier, the bridge was washed out by the current, which caused the whole bridge to collapse into Little River in November 1937. Concrete pilings and some steel can still be found in the river. (Courtesy of the Milam County Historical Commission.)

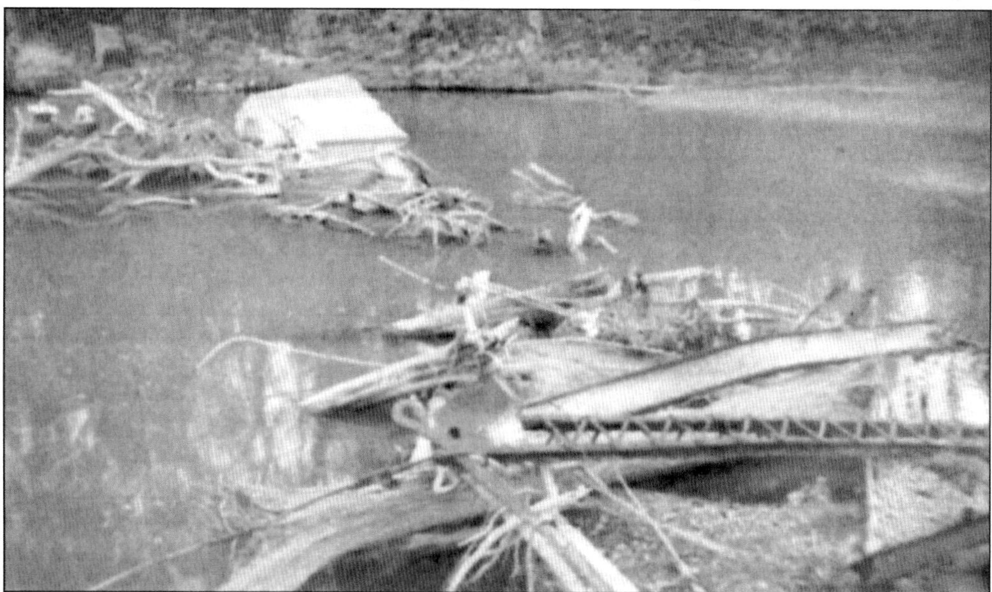

Faubion Bridge is located one mile upstream from Sugarloaf Mountain and served as the main route of transport for the early residents of Milam County on their way to Waco and Temple. The bridge lies on the early historic route called "Ramon's Road." The bridge linked Gause with Port Sullivan, crossing on the Brazos River in route to Nacogdoches and the Old Waco Road. This photograph shows the wreckage still in the Little River, one mile upstream from the County Road 264 cement slab bridge. This was taken in 2002. (Courtesy of David Galbreath.)

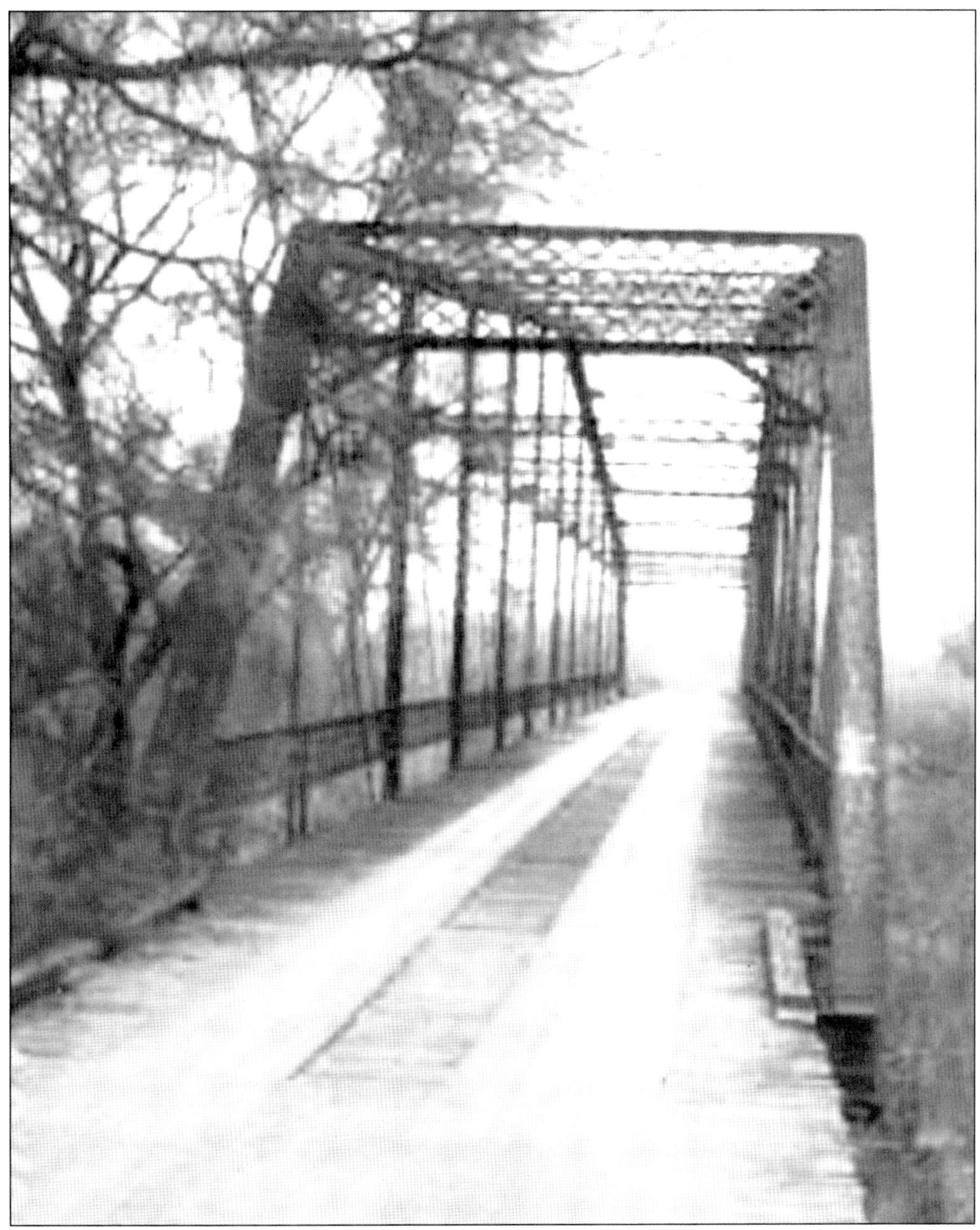

A replacement bridge was built in 1939–1940, using a salvaged span of the old Pitts Bridge over the Brazos on the Caldwell to Bryan road. Located on County Road 264 next to Sugarloaf Mountain, three and a half miles northwest of Gause, this is a Parker through pin truss bridge. Boswell Newton purchased it for $2,500 on September 27, 1939. The Austin Bridge Company of Dallas removed the old bridge and erected a replica of the original bridge at Faubion Crossing near Gause. (Courtesy of David Galbreath.)

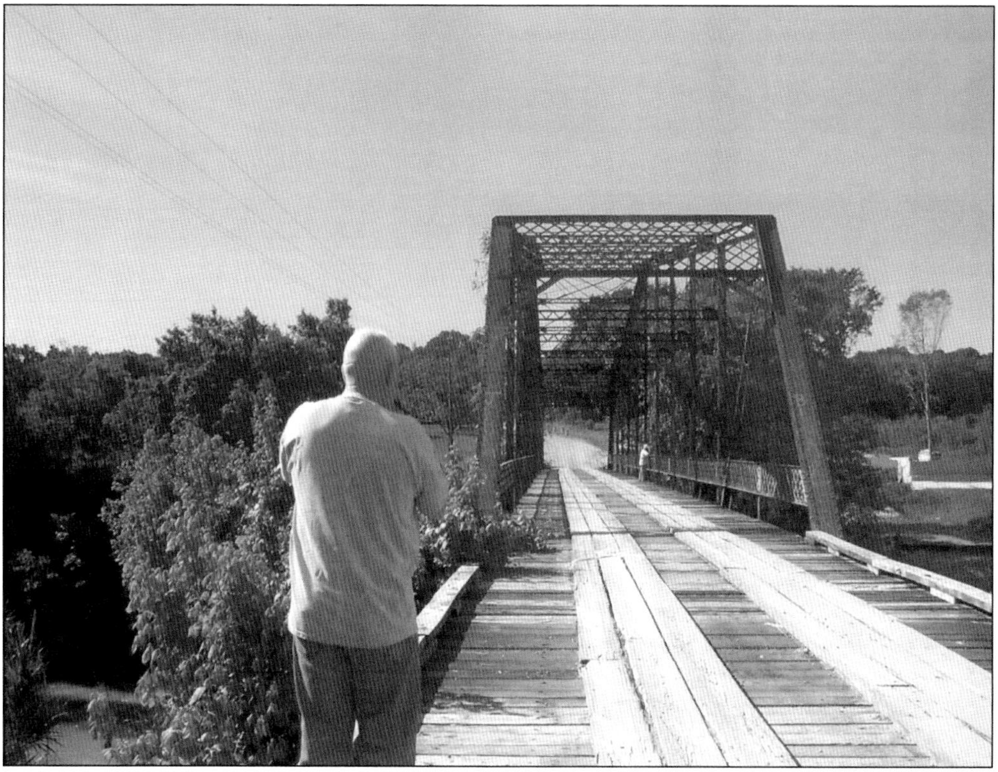

In this view looking south toward Sugarloaf Mountain, David Galbreath takes a picture of Connie Billingsley standing on the bridge and looking upstream. The construction of the cement bridge is upstream to the right. (Courtesy of Randy Billingsley.)

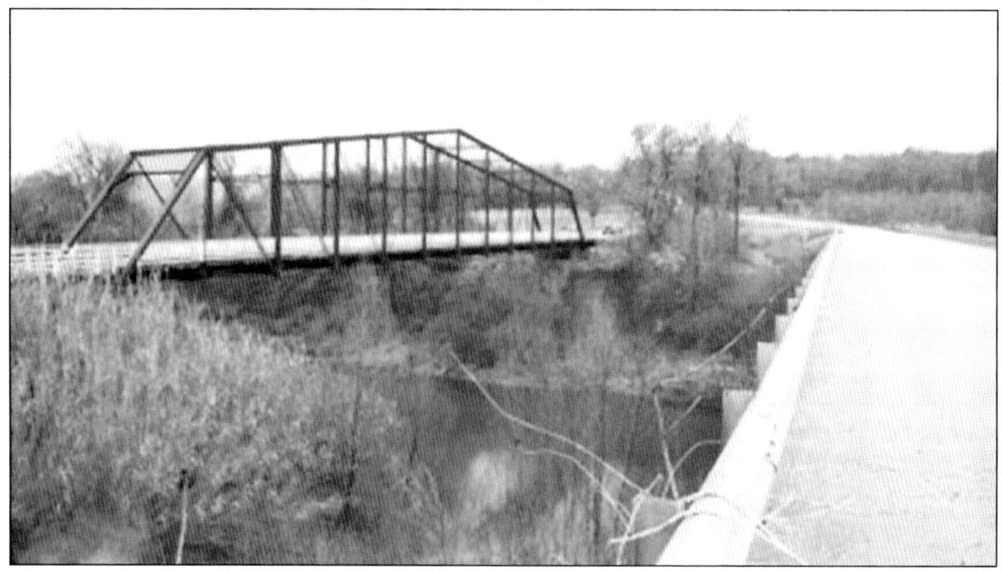

This photograph is of the restored Sugarloaf/Faubion Bridge in 2007 after it was completed as a footbridge. A new cement bridge is to the right and upstream on Little River and County Road 264 north of Gause. The mountain has since been sold, but the bridge is still visited by people from all over. (Courtesy of David Galbreath.)

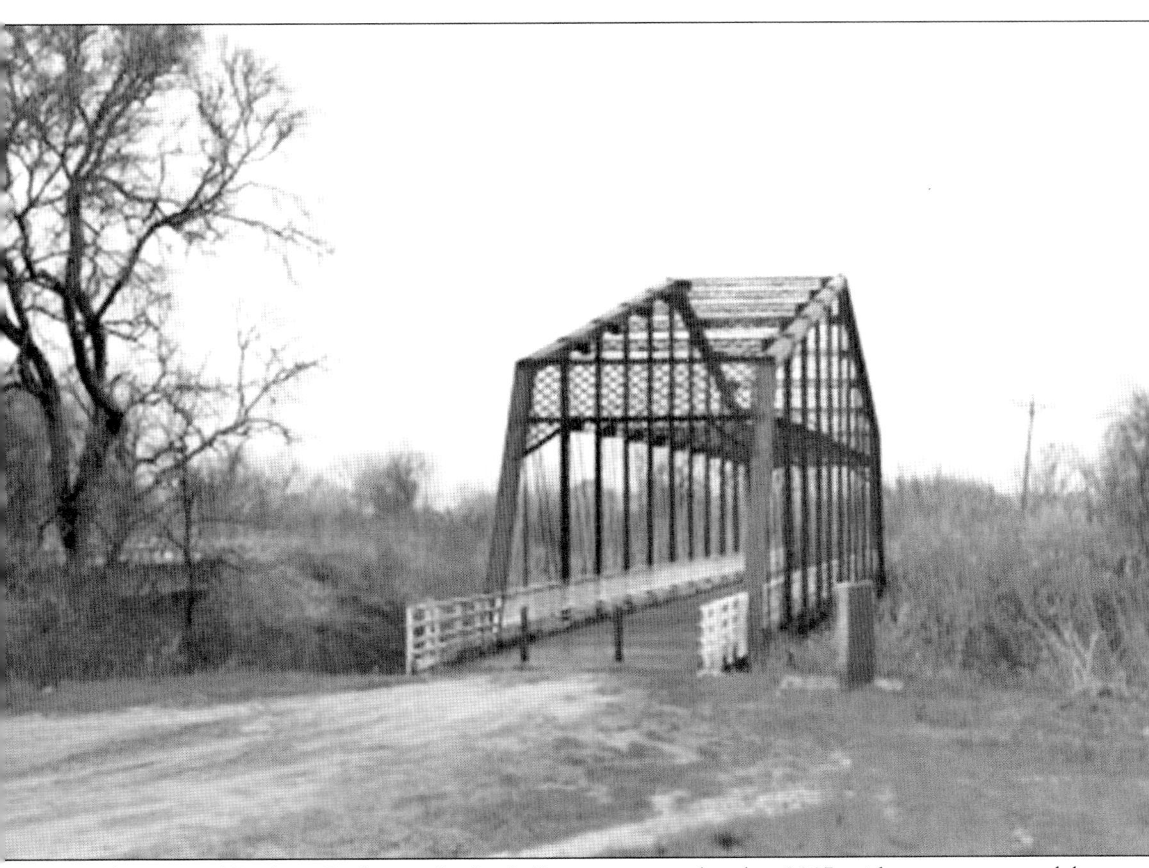

The restored Faubion Bridge on the Little River was completed in 2007, with a new cement slab bridge on the left. Posts were in place to regulate the bridge for pedestrian and bicycle use only and to keep vehicles off for preservation reasons. (Courtesy of David Galbreath.)

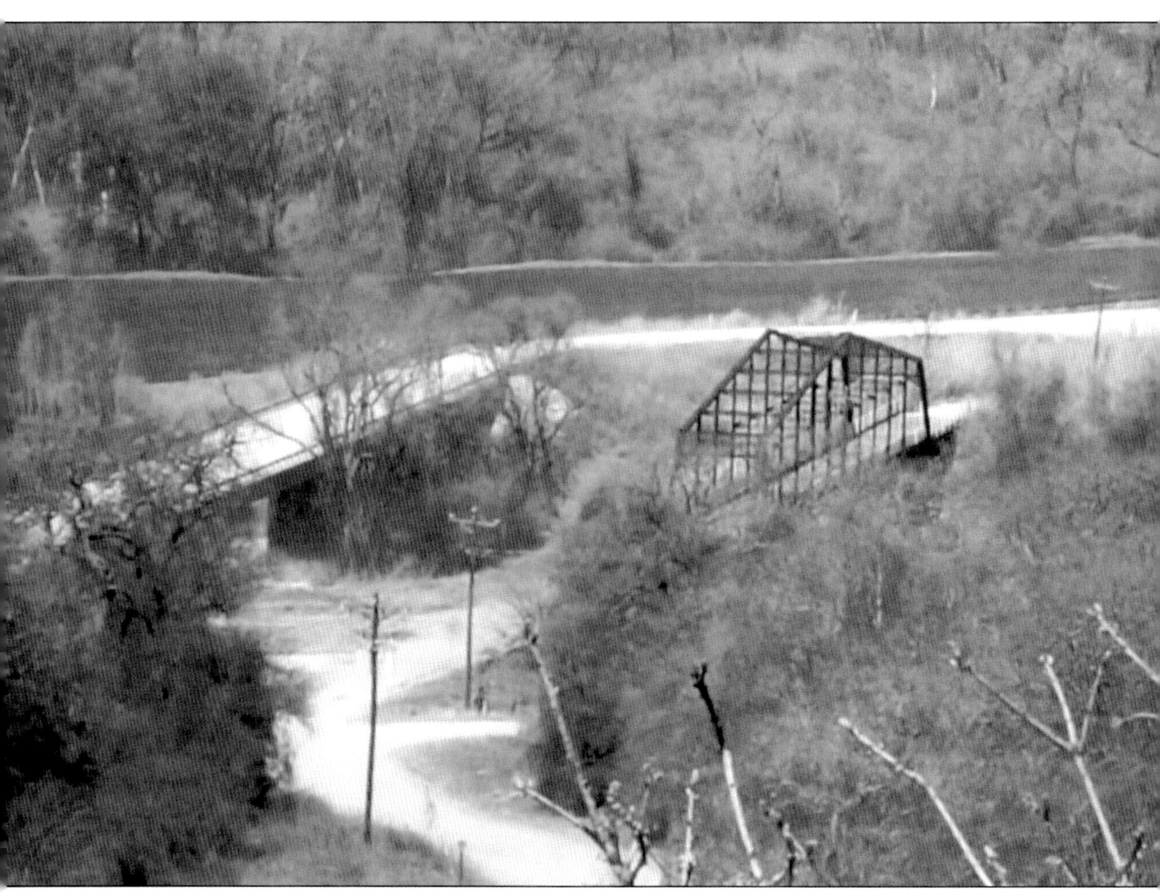

This image shows the new cement bridge and the Faubion/Pitts/Sugarloaf Mountain Bridge from the summit of Sugarloaf Mountain. People would climb the mountain, fish, or swim in the river below. They would walk around the base of the mountain and explore the cave on the north side. One could see the oil tanks at Hearne, the power plant at Alcoa, and Little River downriver as it flowed towards the Brazos. (Courtesy of David Galbreath.)

Two

Historic Bridges, Part Two

Chapter 2 features visits to more of the county's larger bridges. Although none of these bridges are still in use, in most instances, something is visible. The Milam County Historical Commission, now under the leadership of Geri Burnett of Rockdale, has continued its interest in the bridges, manifested by seeking official designations from the Texas Historical Commission. Also, the commission is seeking to find out which bridges are eligible for listing in the nation's most prestigious register, the National Register of Historic Places. Once the eligibility is determined, an effort will be made to complete these listings.

Bridges included in this chapter are Galbreath Crossing Bridge, Six Mile Creek Bridge, Elm Creek Bridge (two bridges, both named Elm Creek), and Donahoe Bridge. The reason for the creation of each bridge is evident. Milam County is a large county, and its successful development depended on residents being able to move both themselves and good to necessary places, such as markets. In addition to the bridges listed here, there are within the county innumerable fords. Through the years, county commissioners have worked with residents to make the crossings accessible and safe.

Note that Galbreath Crossing Bridge has been relocated to Bridge Park in Rockdale. Be sure to read more about that in chapter 6.

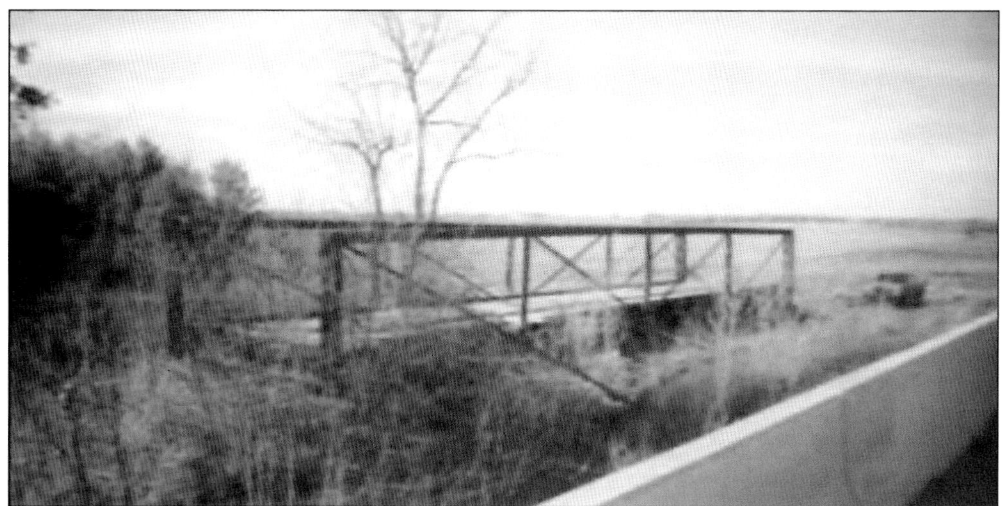

This is a photograph of Galbreath Bridge on the northwest bank of Big Elm Creek on County Road 240. For years, it was left for a foot bridge when it was bypassed in 1986. It was moved to the northwest bank in 1994. During the next 20 years, it became overgrown with brush and weeds. In 1986, Milam County judge Gene F. Blake announced to the county Commissioners Court for Precinct 2, that the bridge structure at Big Elm Creek was slated to be replaced by the State Department of Highways and Public Transportation. The new bridge was built on a new alignment east of the present bridge, and the existing bridge was to be left in place. Judge Blake presented a resolution to the State Preservation Office for the existing structure's eligibility for inclusion with the National Registry of Historic Places. (Courtesy of David Galbreath.)

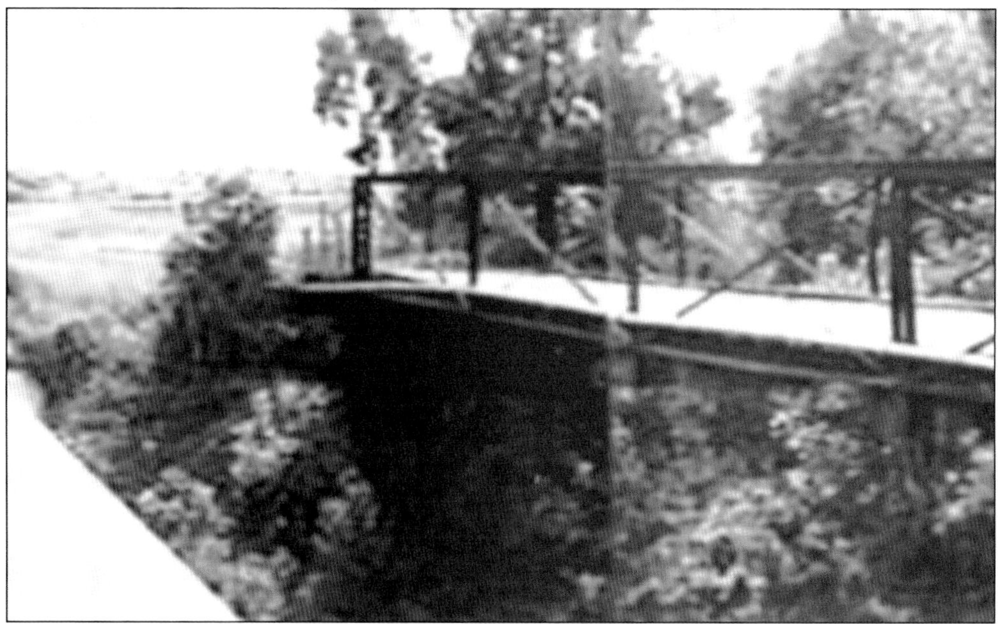

The Galbreath Crossing Bridge of Big Elm Creek was originally built in 1876 on County Road 240, northeast of Cameron Airport on the Silver City and Maysfield Road. The bridge linked Milam County with the "Old Waco to Washington Road" and gave access to Calvert and Hearne. The bridge was on a route to Cameron with a link to Maysfield and the extended area on to Brazos Crossing at Port Sullivan. (Courtesy of David Galbreath.)

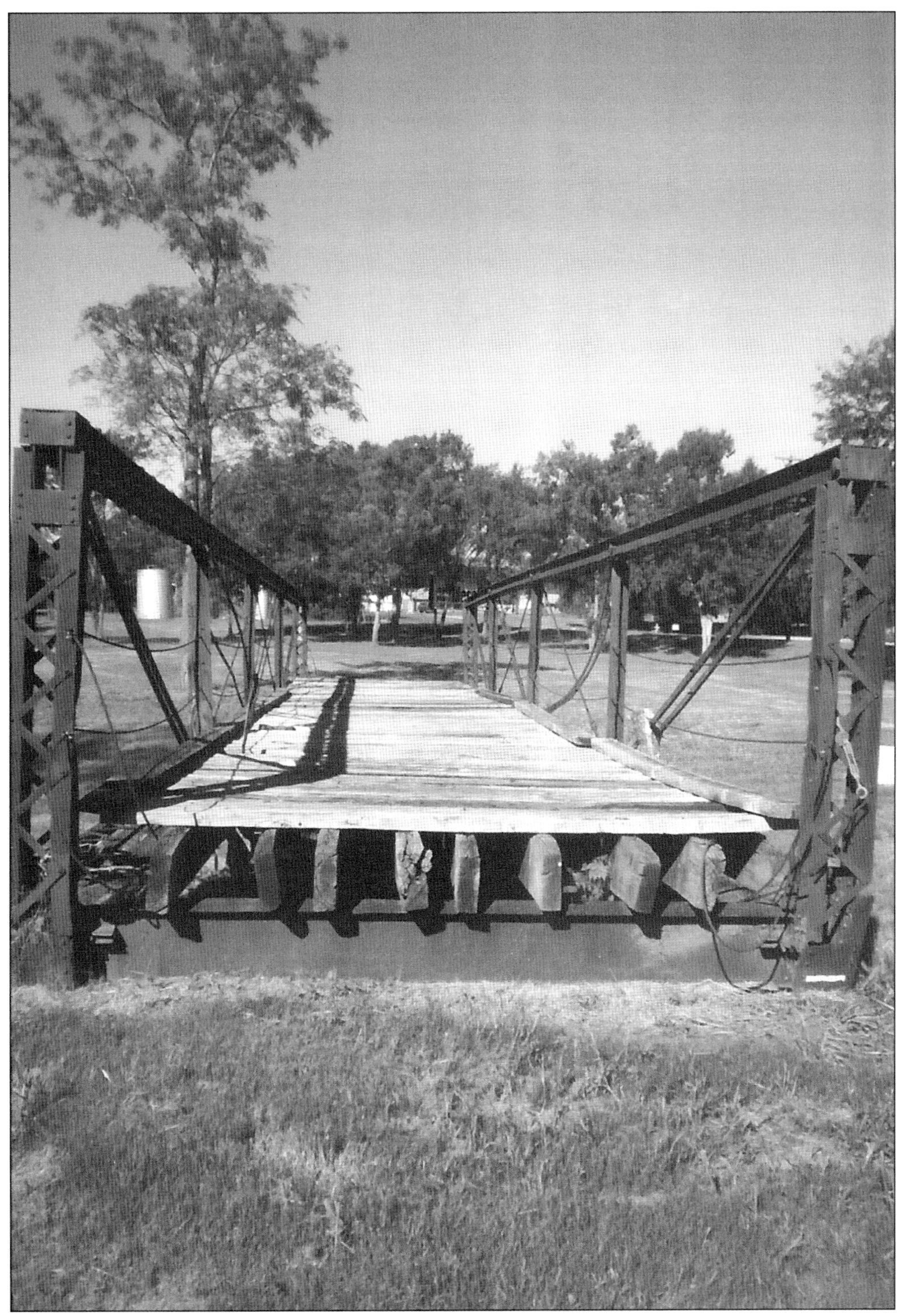

Galbreath Crossing Bridge is shown at Bridge Park in Rockdale, Texas. It was built in 1912 by the El Paso Bridge Company, with the steel bedstead truss and a creosoted wood deck. It spanned a total of 115 feet in length across the creek. (Courtesy of David Galbreath.)

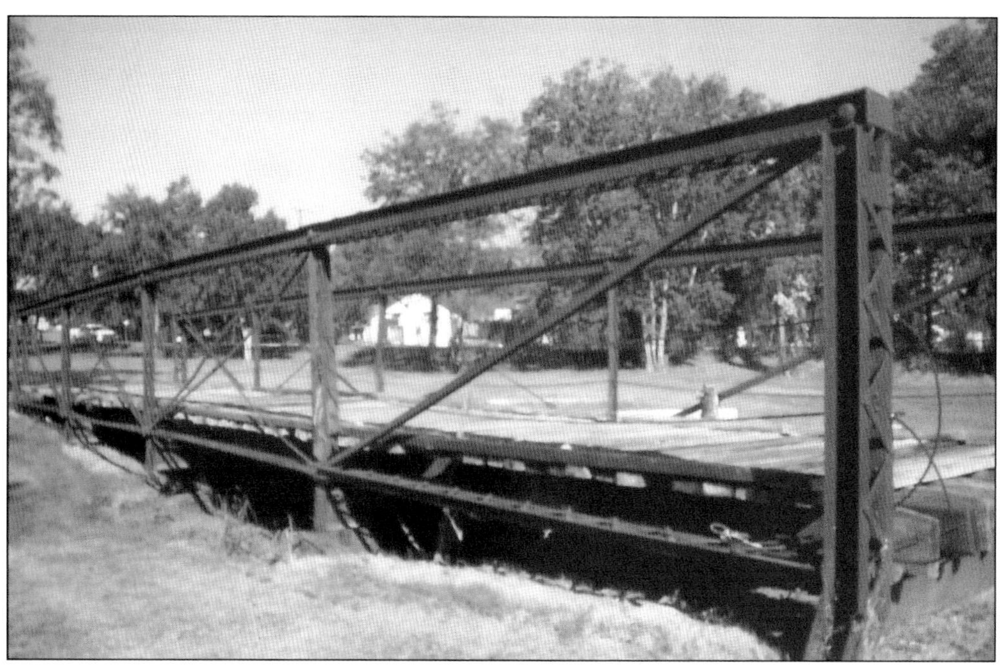

Both photographs show Galbreath Crossing Bridge at Bridge Park. Cement piers have been created to rest the bridge on to make it more stable. The timbers need to be replaced and repairs to the metal structure are needed. (Both, courtesy of David Galbreath.)

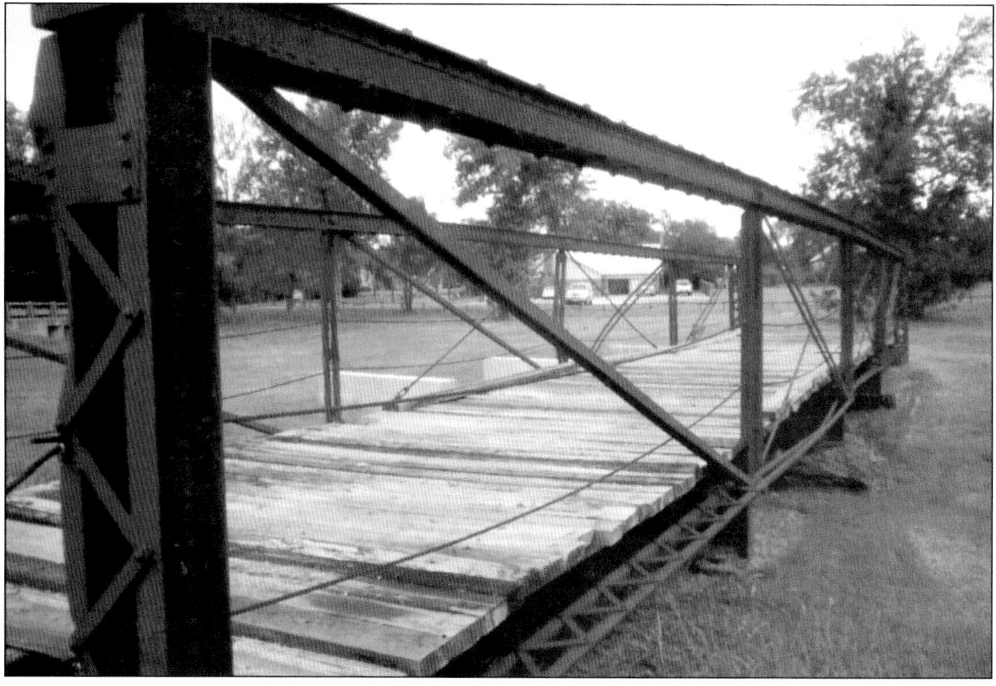

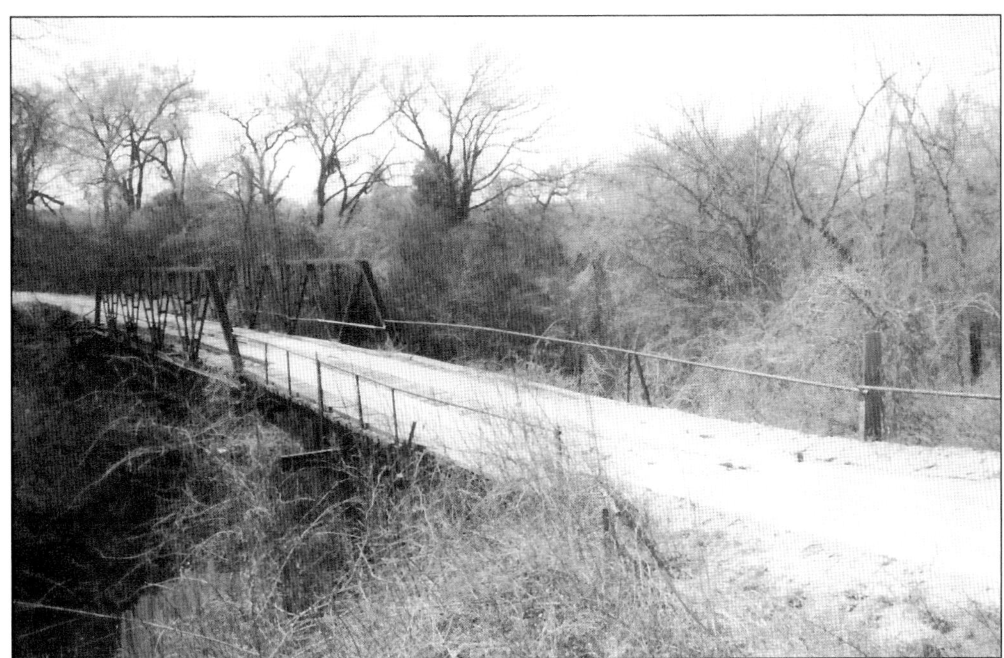

Six Mile Creek Bridge was located southeast of Gause on County Road 358. It was 90 feet in length and was a Warren Penn truss design. (Courtesy of Dolores Mode.)

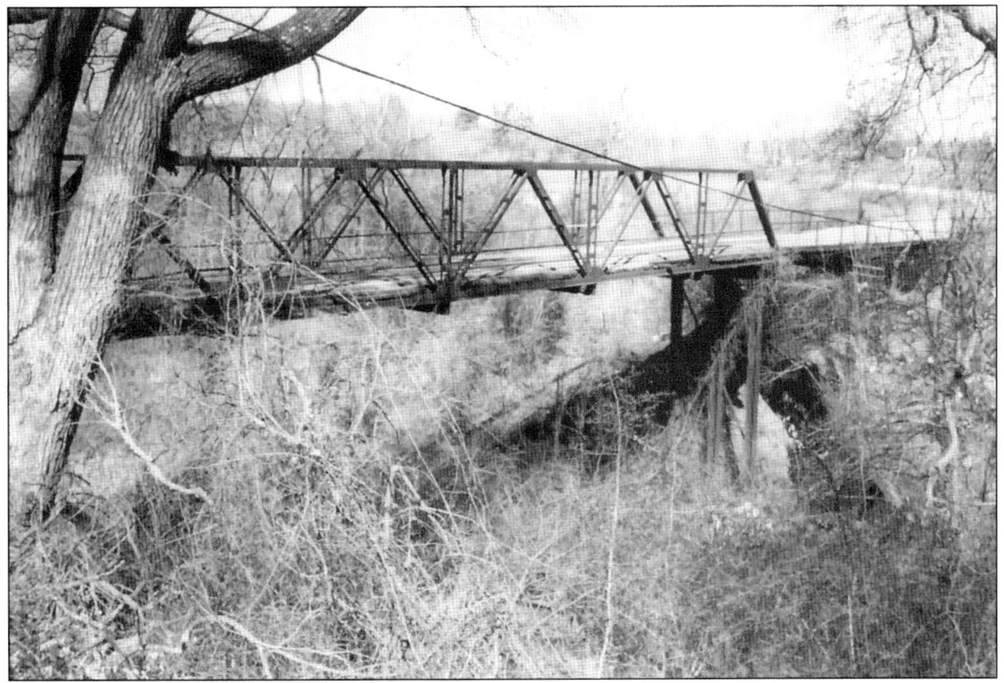

Six Mile Creek rises six miles south of Gause in eastern Milam County and runs east for eight miles to its mouth on the Brazos River, a mile north of Milam-Burleson County line. The creek was named for its distance from old Nashville, the early capital of Robertson's Colony. Six Mile Creek Bridge collapsed in December 1997 when an oil field tanker attempted to cross it. The driver was uninjured, but the bridge was destroyed. (Courtesy of Dolores Mode.)

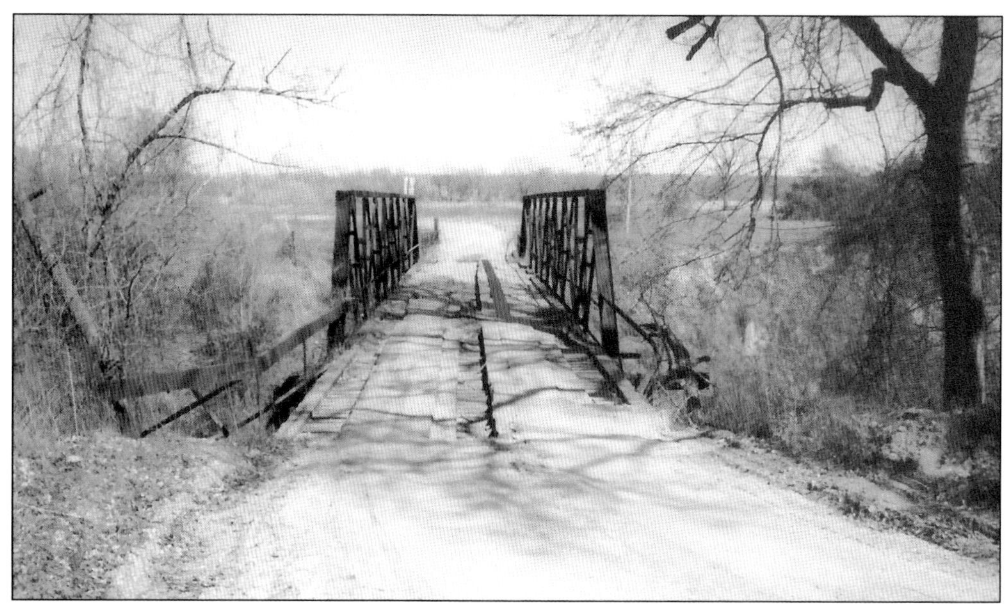

Six Mile Creek has had three bridges constructed over it; they were not massive structures and did not last long. In 1903, the Six Mile community had one teacher for 13 white students and one teacher for 29 black students. By the 1940s, no evidence of the community appeared on county highway maps. (Courtesy of Dolores Mode.)

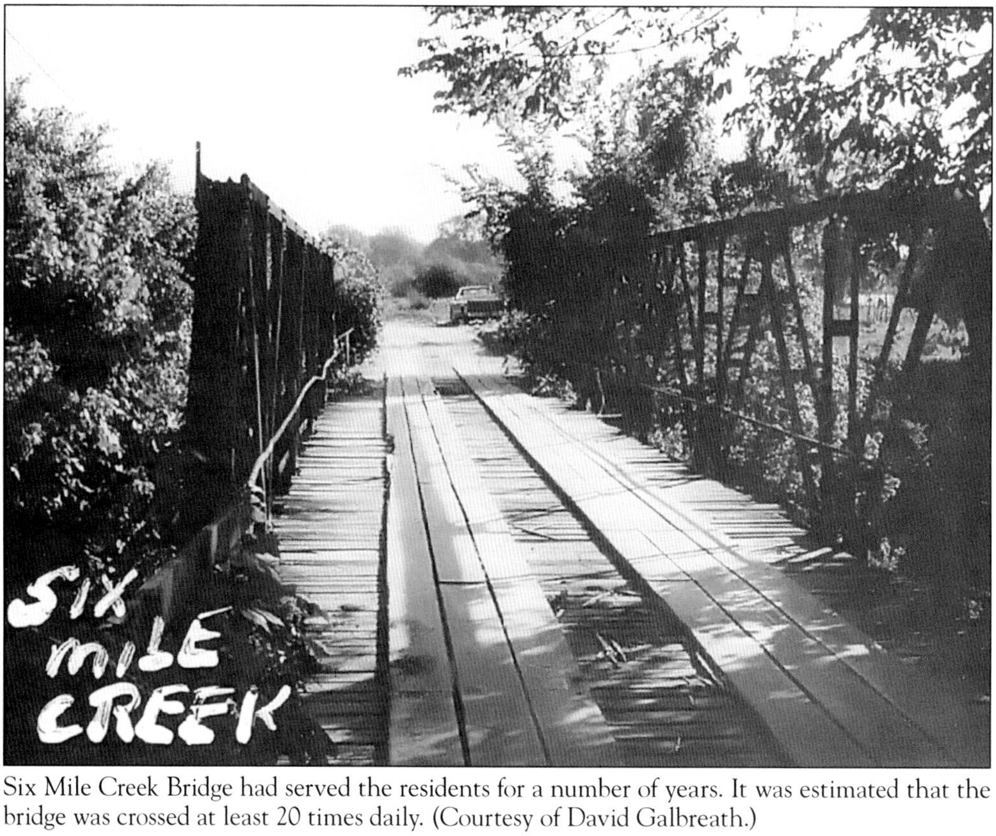

Six Mile Creek Bridge had served the residents for a number of years. It was estimated that the bridge was crossed at least 20 times daily. (Courtesy of David Galbreath.)

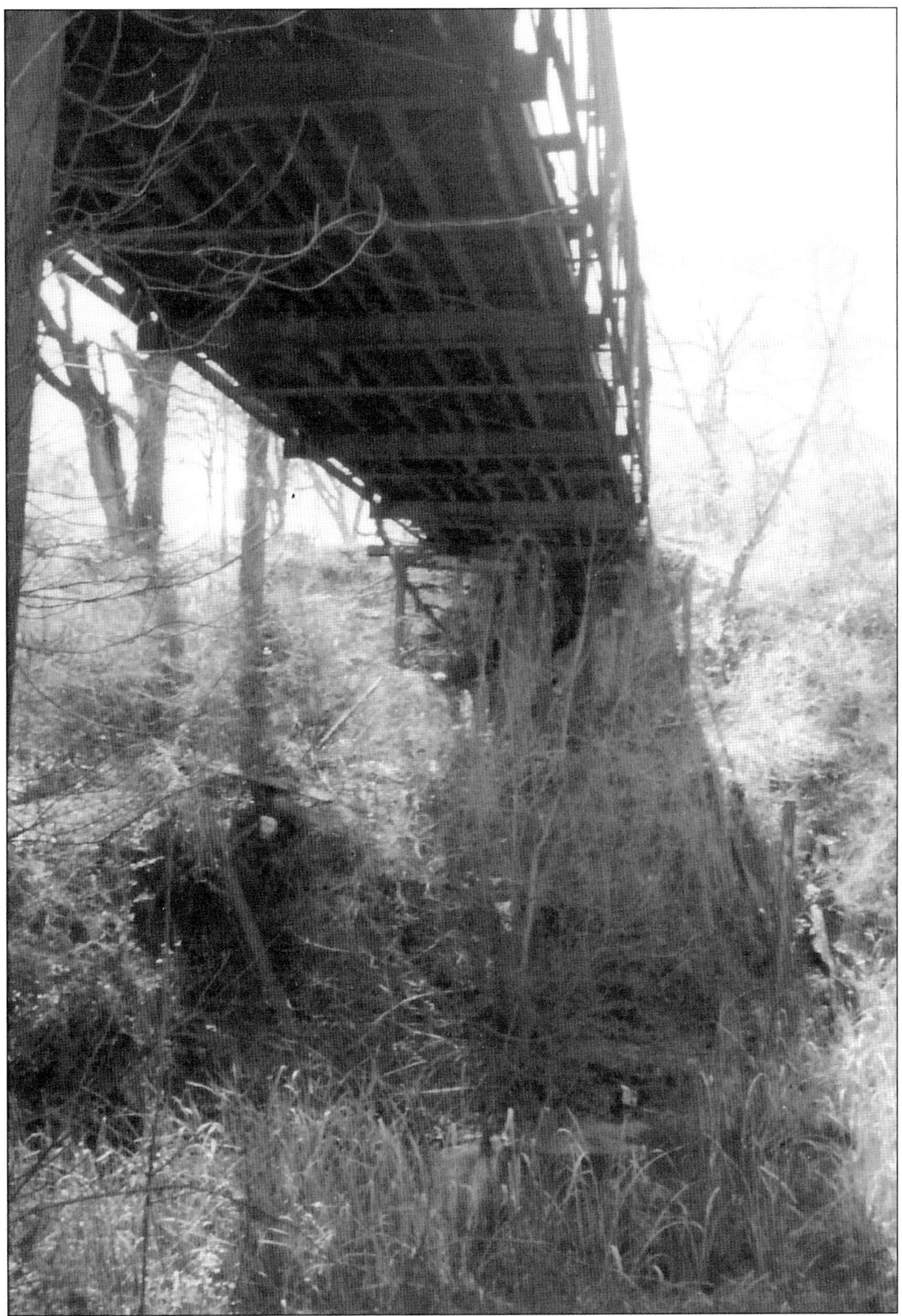

This photograph of the underside of Six Mile Creek Bridge shows how the bridge was made stable. The public needed this bridge in order to cross to the other side of the creek. (Courtesy of Dolores Mode.)

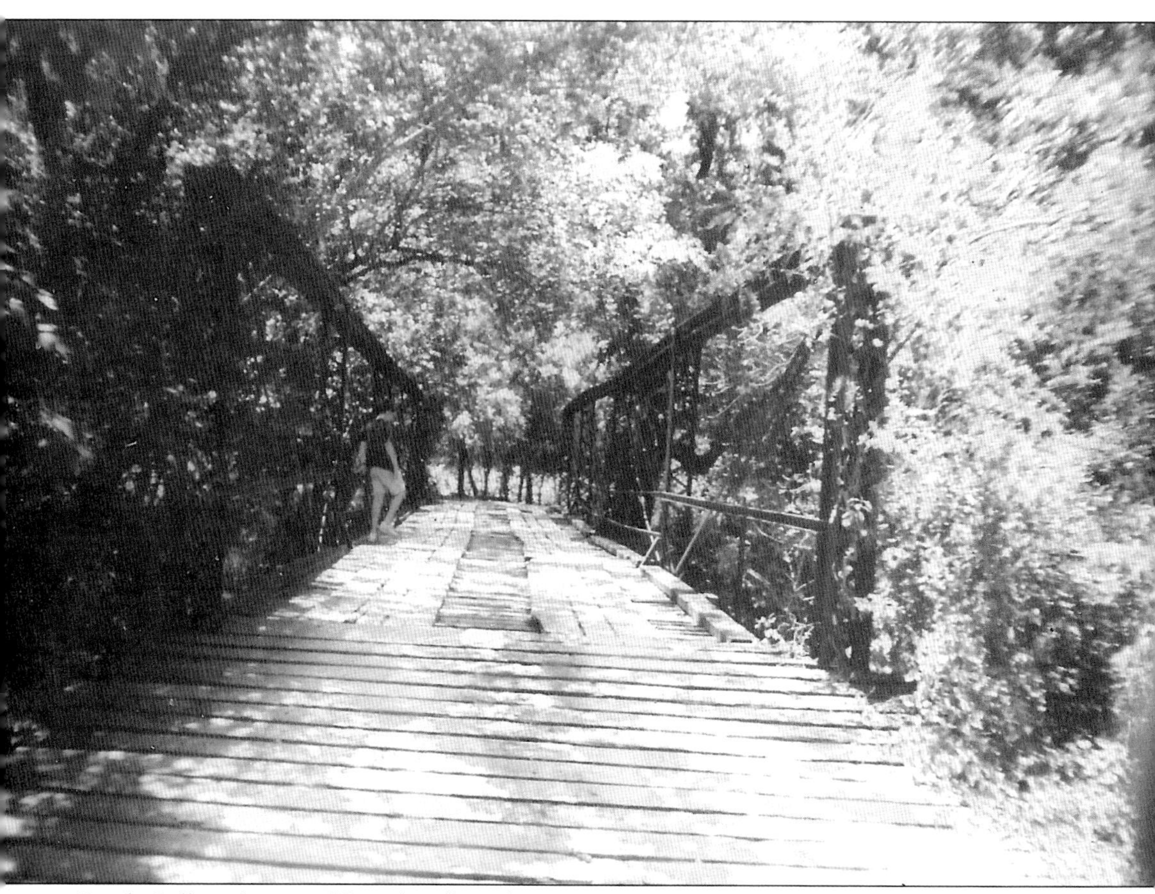

Derek Galbreath, son of David Galbreath, is on the Elm Creek Bridge. The bridge was built northwest of Cameron on County Road 119 in 1898. This view looks westward. In Texas history, the elm tree became famous in the battle of January 7, 1837, in Milam County on the creek that now bears the name Elm Creek. Capt. George B. Erath led a small force of Texas Rangers, and he described these trees as "friendly trees which sheltered my men from the charging Indians." (Courtesy of David Galbreath.)

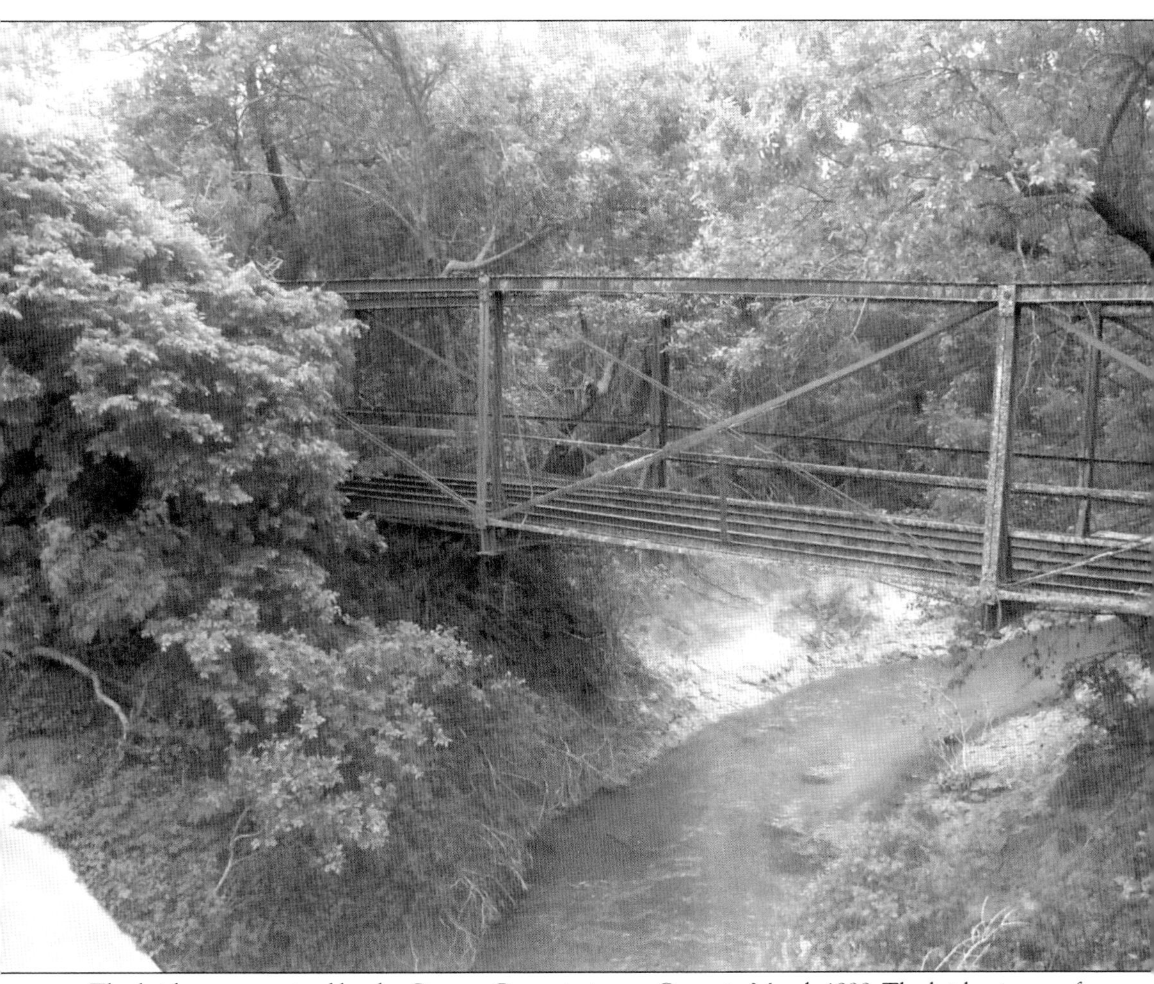

The bridge was received by the County Commissioners Court in March 1899. The bridge is one of six bedstead truss type bridges surviving in Texas, and it is the longest in the United States. The Elm Creek Bridge today is just the metal frame of the bedstead bridge and is no longer operational for traffic. The bridge has retained its original materials and has not been substantially altered. The deck is deteriorated, but the steel framing is intact. (Courtesy of David Galbreath.)

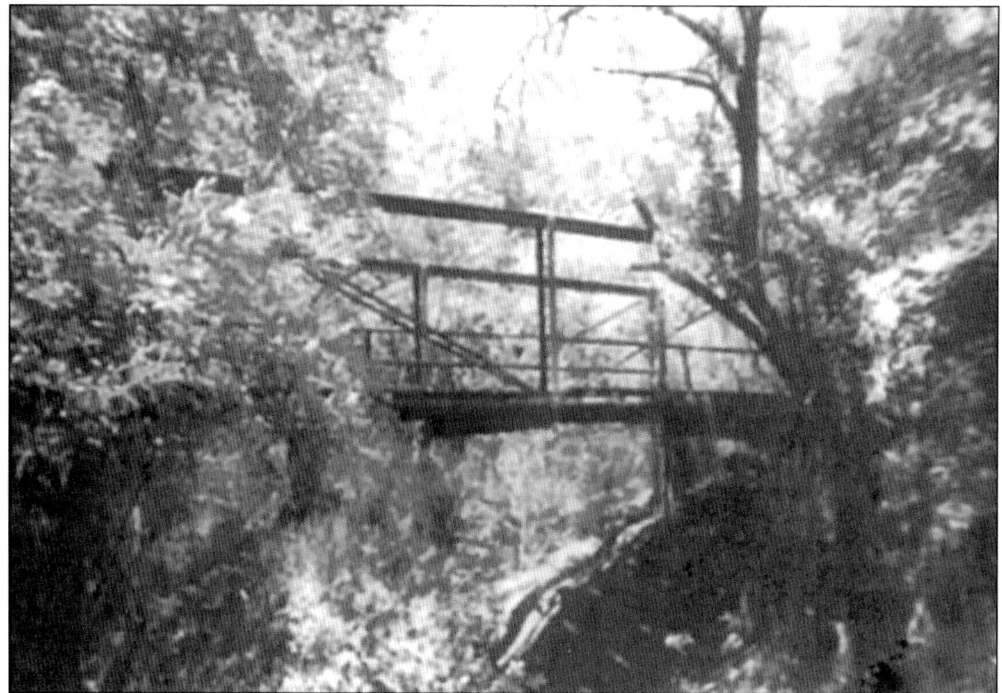

Taken in 1986, this is a view upstream from Elm Creek Bridge on County Road 119. Elm Creek cuts across Milam County in the northwest quadrant, east of Texas 36/US190 and north of Buckholts. The land on Elm Creek was desired by early settlers as a water source and used for agriculture and for grazing cattle. When the names of the streams were established before 1835, North and South Elm Creeks came together near the Marek School to form Big Elm Creek, sometimes called "Deep Ellum." It flows southeast to join the Little River down toward the Brazos, and many small communities formed along it, including Cameron. (Courtesy of David and Derek Galbreath.)

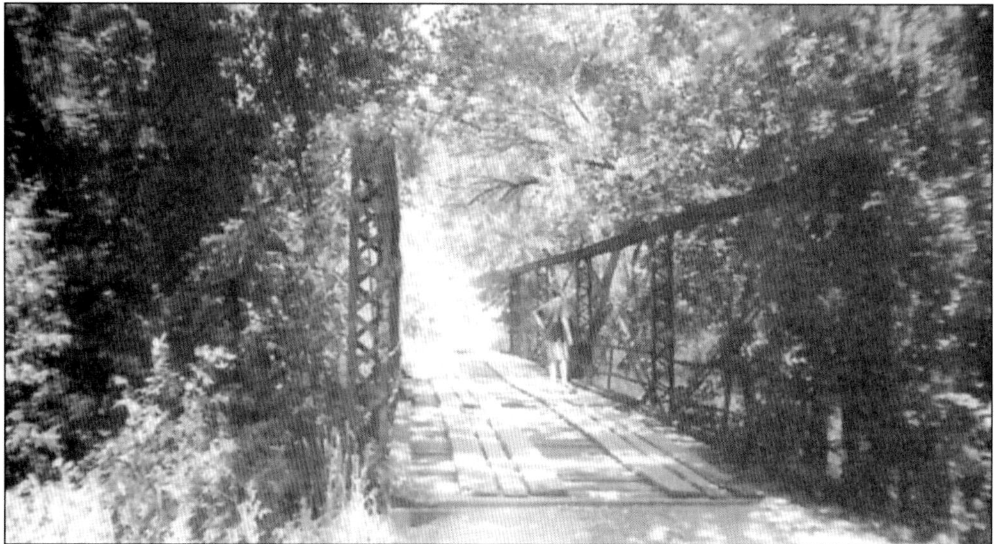

The Elm Creek bedstead truss bridge was 91 feet in total length, with an 80-foot span length and 12-foot pavement width. It was erected in 1898 between Cameron and Rogers by the Rhoades Bridge Company. (Courtesy of David Galbreath.)

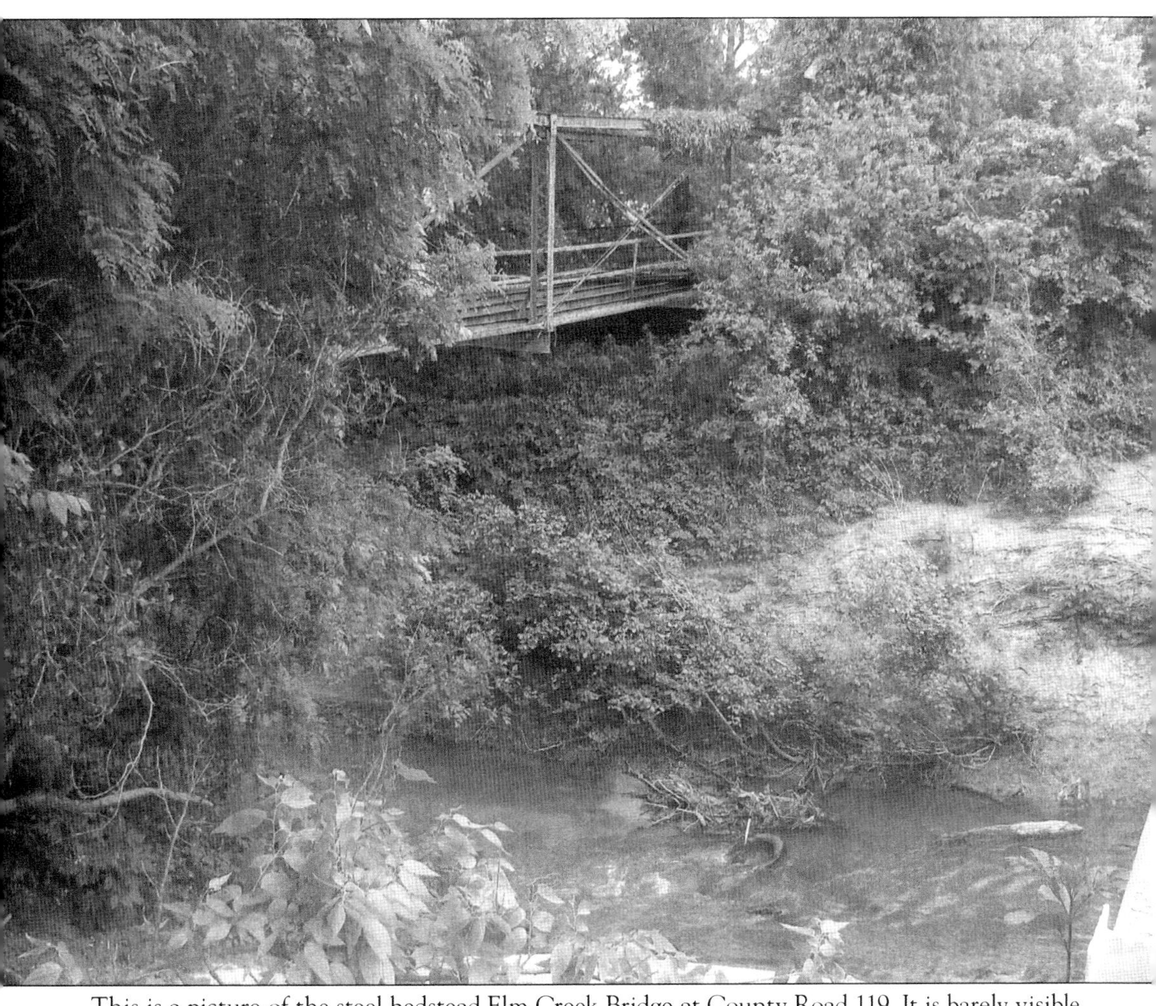

This is a picture of the steel bedstead Elm Creek Bridge at County Road 119. It is barely visible through the trees, with Elm Creek flowing below. The photograph was taken from the cement slab bridge in the spring of 2016. (Courtesy of David Galbreath.)

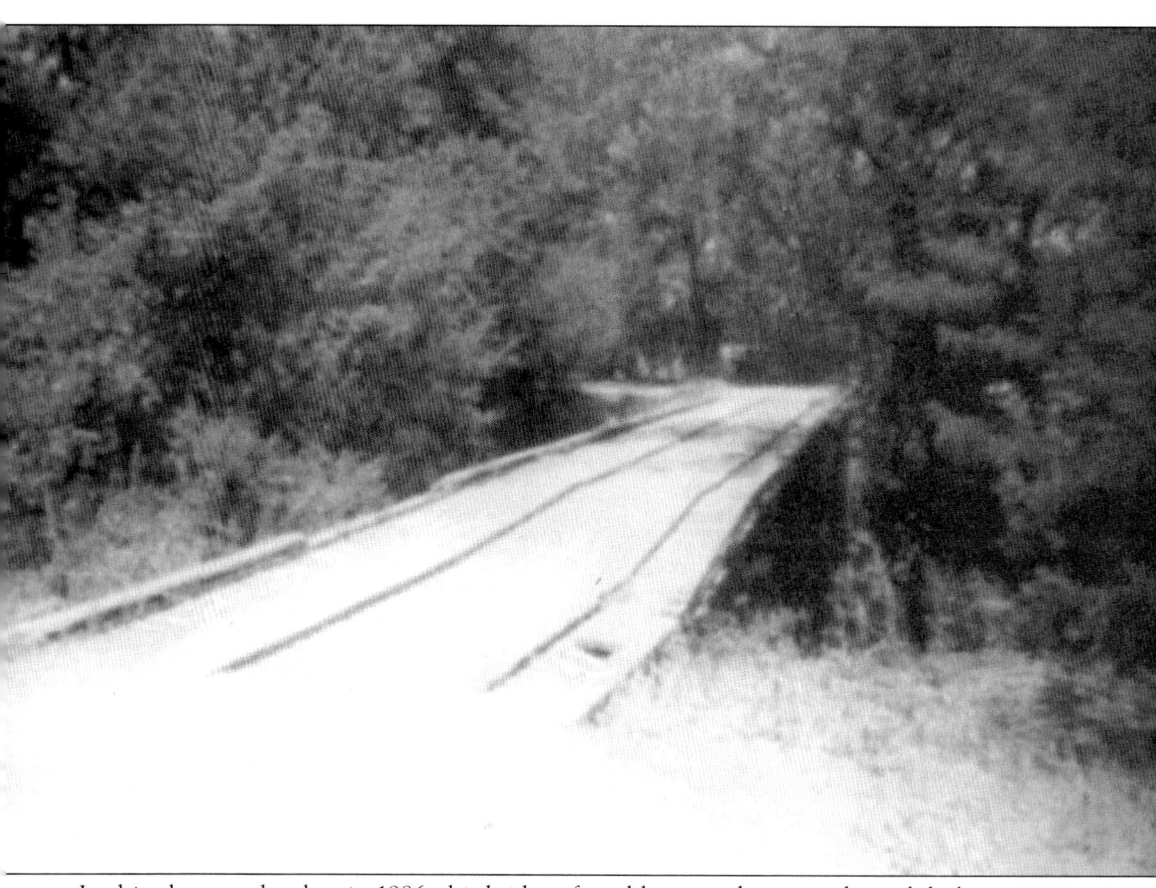

In this photograph taken in 1986, this bridge of steel bents and creosoted wood decking spans over South Elm Creek in northwest Milam County. The structure is 92 feet in total length with 12.5-foot-wide pavement and a 76-foot span length. The Elm Creek Bridge is located west of Buckholts and about two miles east of Farm to Market 1915 on County Road 120. The bridge marks the route between Cameron and Buckholts and is part of the enhanced traffic facilities on to Roger that link to Temple and Taylor. (Courtesy of David Galbreath.)

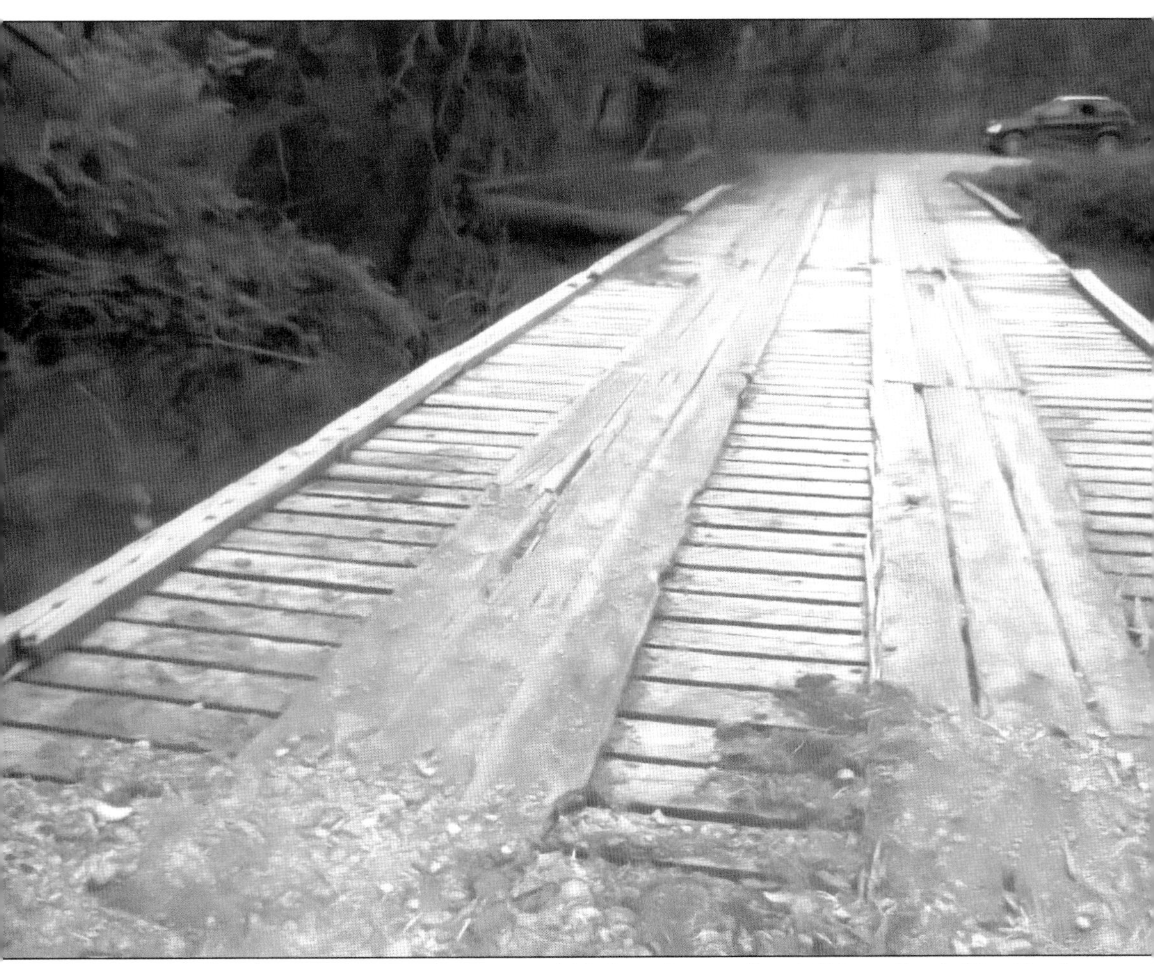

This photograph was taken in the spring of 2016 on County Road 120 on South Elm Creek, with a view looking toward the southeast. The bridge was rebuilt in 1987. Buckholts is a rural community of cattle and farming residents who depended on the bridges along Elm Creek to hold up against the hazardous conditions so that they could have access to the hub of economic trade in the area. Without access across the creek, many residents went without means for providing for their families. (Courtesy of David Galbreath.)

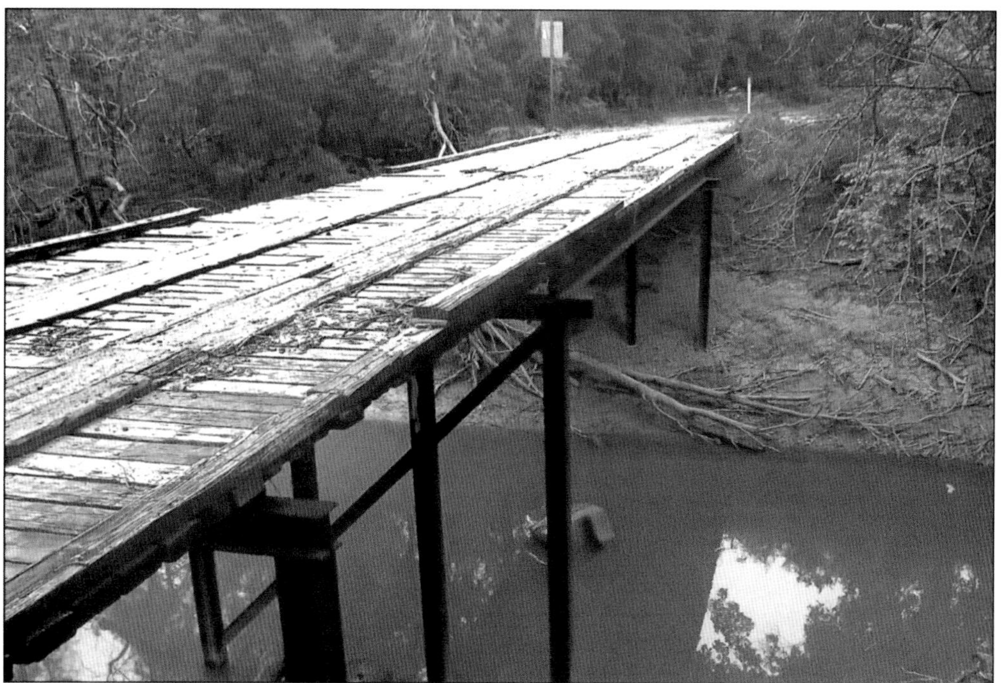

This photograph shows a view of Elm Creek Bridge on County Road 120 across Elm Creek. It has steel bents and wooden decking. This view looks from the southwest corner of the creek towards Cameron. (Courtesy of David Galbreath.)

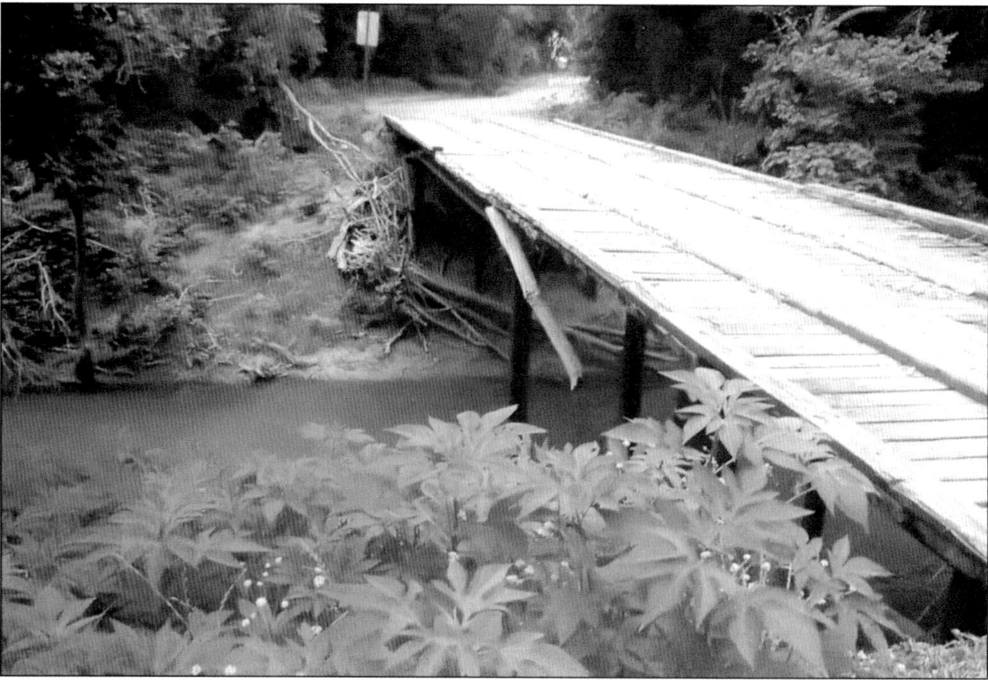

The view in this photograph, taken in the spring of 2016, looks from the east of County Road 120. This picture of Elm Creek Bridge shows it in need of repair. The side railing has driftwood underneath the steel bent piers in the creek bed. (Courtesy of David Galbreath.)

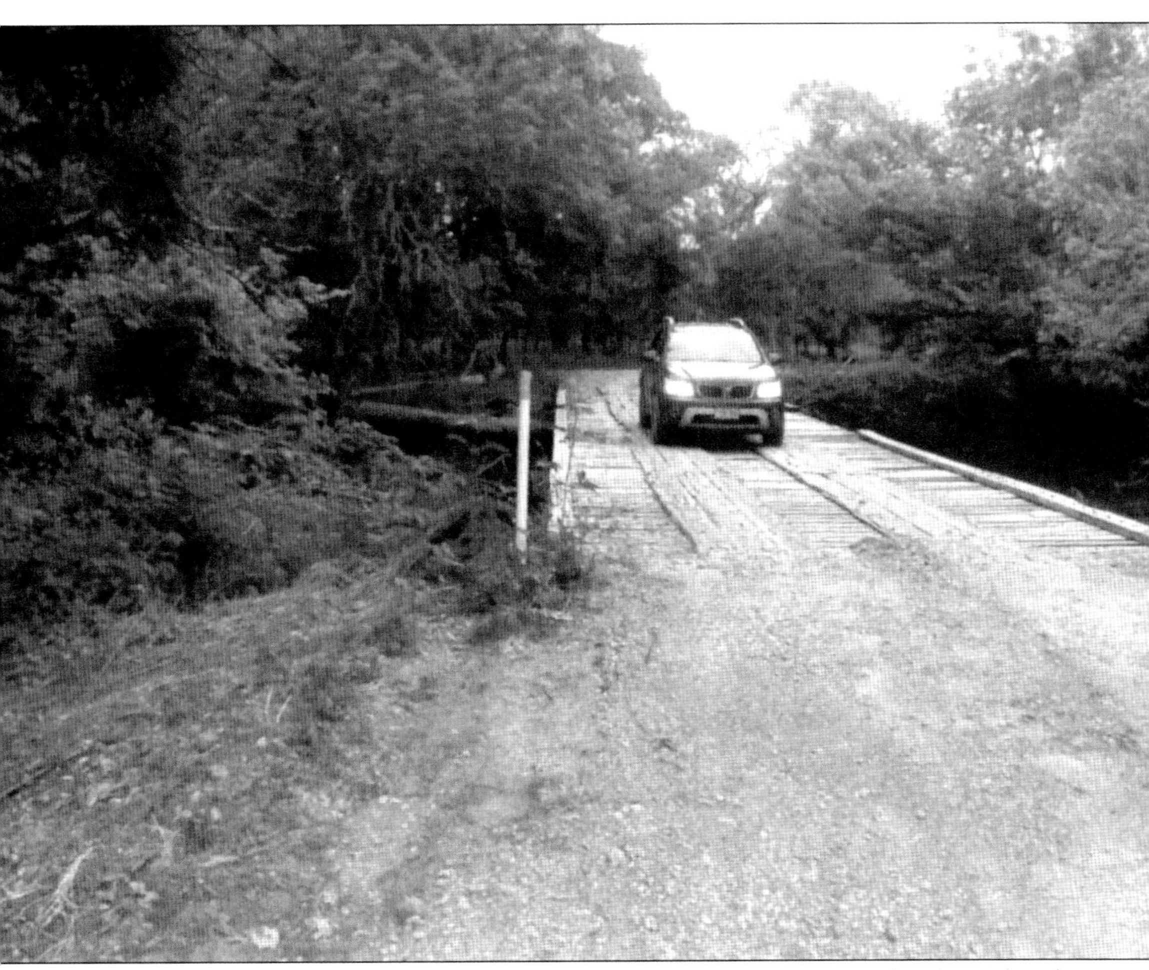

Taken in April 2016, this is a photograph of a vehicle crossing the bridge. This shows that the bridge is still passable for oncoming vehicles and used in the daily lives of residents in this rural area. (Courtesy of David Galbreath.)

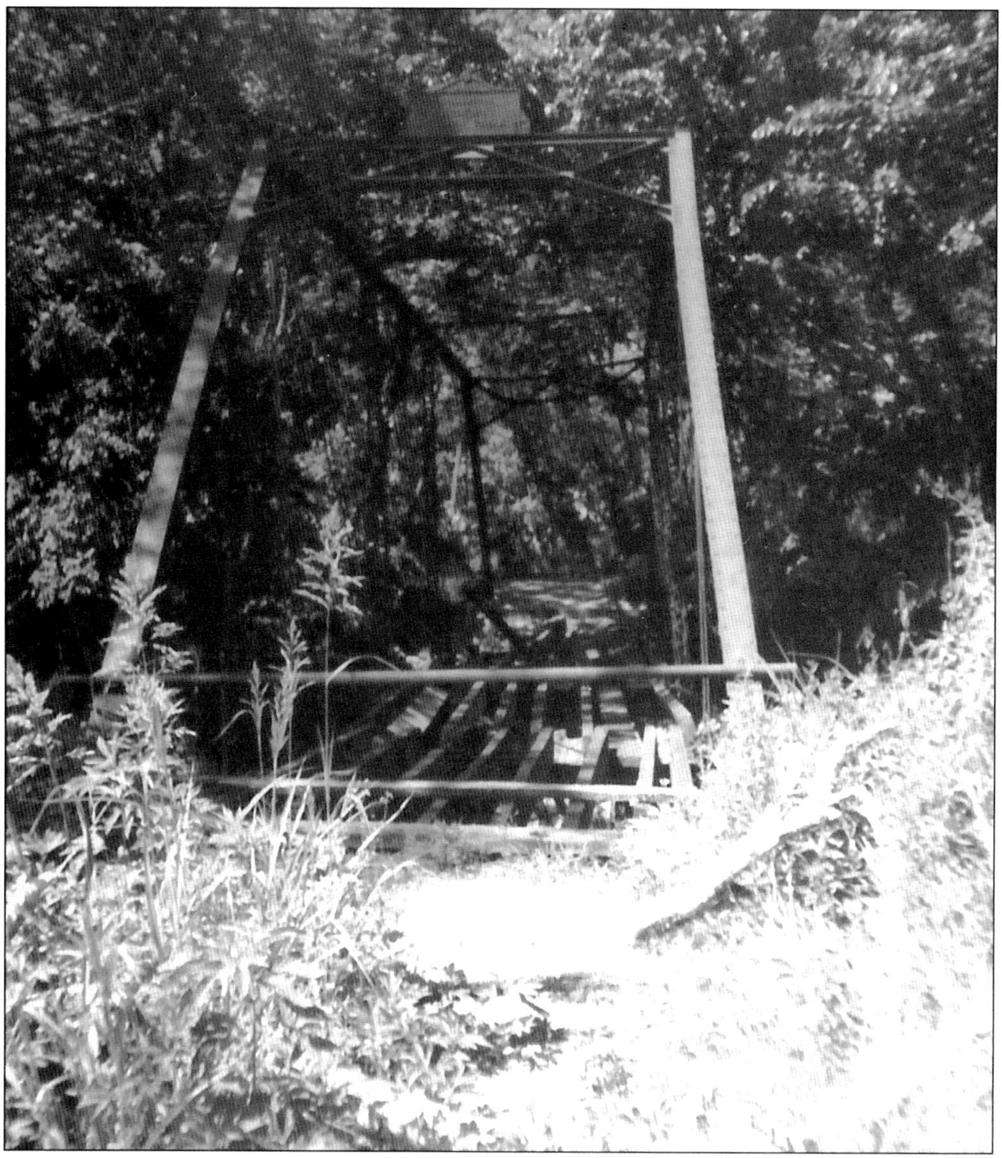

This is a Pratt through pinned bridge over Donahoe Creek south of Milam County Road 402, northeast of Davilla. It was erected in 1908 by the C.Q. Horton Company of Austin. The bridge allowed residents easy access to the town of Rogers, which was a major point for the Gulf, Colorado & Santa Fe Railway. (Courtesy of David Galbreath.)

This photograph of Donahoe Creek Bridge was taken in 1986 and shows the bridge plaque for builders C.Q. Horton of Austin, Texas, in 1908. It was a Pratt through truss–designed bridge with a wooden deck. The span was replaced in 2001 with a cement slab bridge. (Courtesy of David Galbreath.)

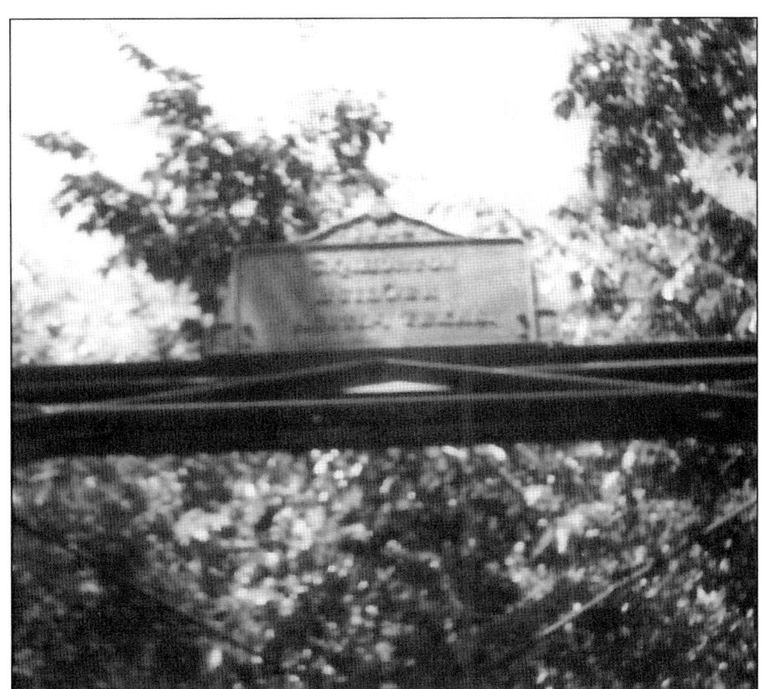

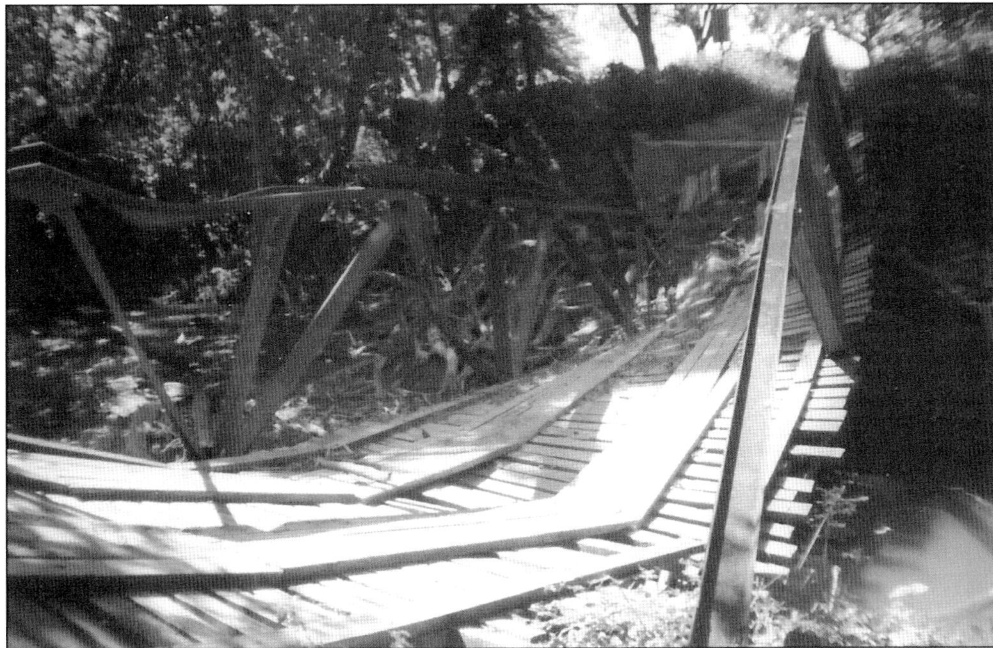

This photograph was taken in 1986 after the Donahoe Creek Bridge off Milam County Road 403 had collapsed into the creek bed possibly due to natural elements and erosion. The bridge was a Warren Pony truss bridge with cement piers at each end. After 1986, this bridge was removed and only the cement piers remained. By the middle of the 1990s, the county road was closed and made private with a gate. (Courtesy of David Galbreath.)

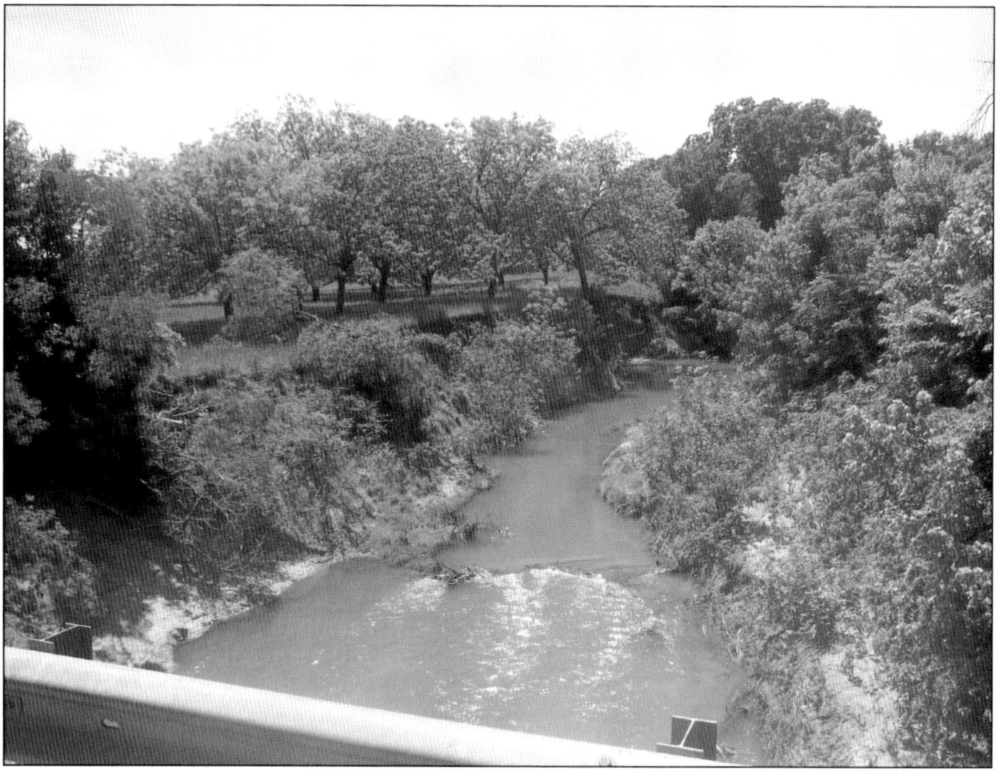

This is a view of Donahoe Creek flowing under the cement slab bridge on County Road 402 in northwest Milam County in 2015. (Courtesy of David Galbreath.)

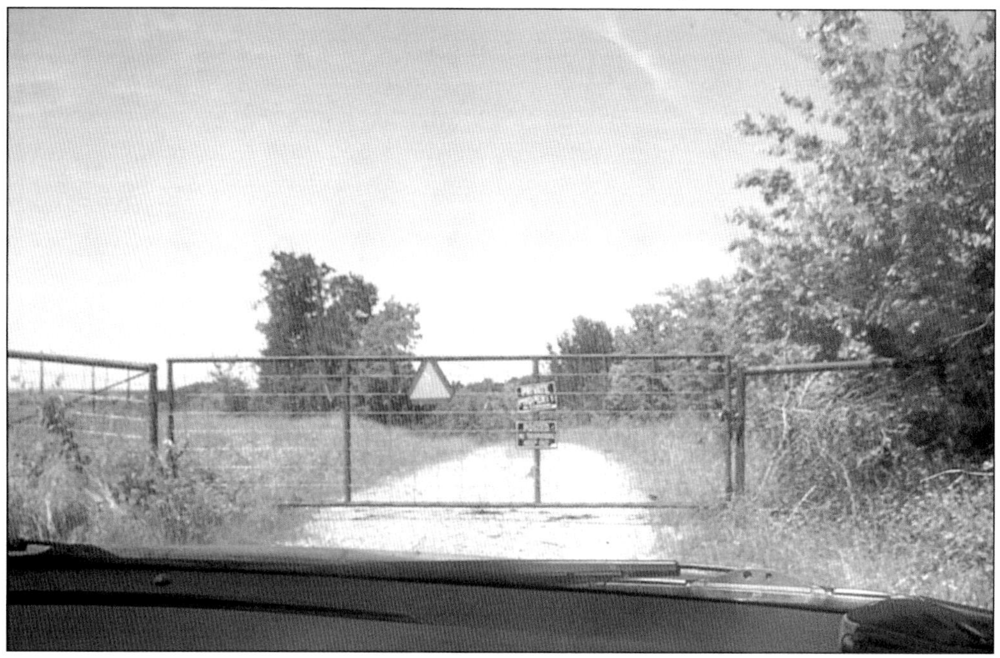

County Road 403 was changed to Private Road 5403 and closed off with a gate across the road. This picture was taken in 2015. (Courtesy of David Galbreath.)

Three

Historic Bridges, Part Three

Chapter 3 concludes a discussion of the major bridges in Milam County. The term *major* is used in the context meaning truss bridges of significant size or with geographic locations giving them a special importance to Milam County. This chapter includes information about Hog Creek Bridge, Sheckles Bridge, Henderson Brushy Creek Bridge, McCown Bridge, and the Little River Bridge. Two of these spans have been moved. Sheckles Bridge is now located in Rockdale at the Bridge Park, and Henderson Bridge was moved to Rockdale's Skate Park several years ago. There are also metal truss bridges, which represent a dwindling historical resource—not only in Texas, but throughout the nation.

It is interesting to note that despite the fact that some of these bridges are totally gone, there remains some means of getting across rivers and streams. As these bridges were replaced—with the exception of Worley Bridge, which was restored instead of being rebuilt—the newer structures were very plain. There is no ornate ironwork and, in most instances, the replacement is merely a cement slab bridge that serves the same purpose as the larger, more elaborate structure. Why then are these old structures described and even commemorated? Perhaps this is simply a salute by historians to a way of life and values now decades old. It is possible that by recognizing these values that the doors of history will open to even greater heights.

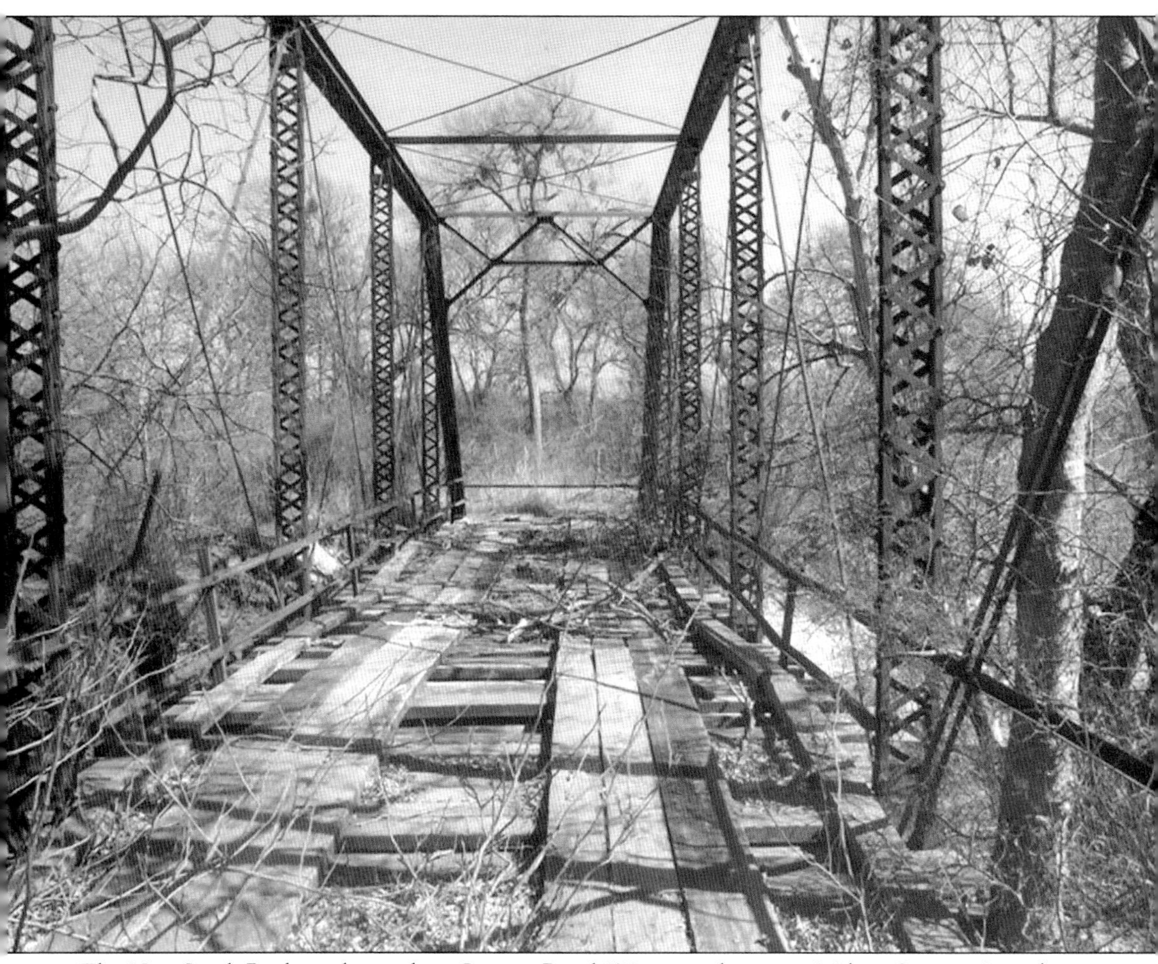

The Hog Creek Bridge is located on County Road 278 in northeastern Milam County. It is also 1.75 miles southwest of Farm to Market Road 2027. (Courtesy of David Galbreath.)

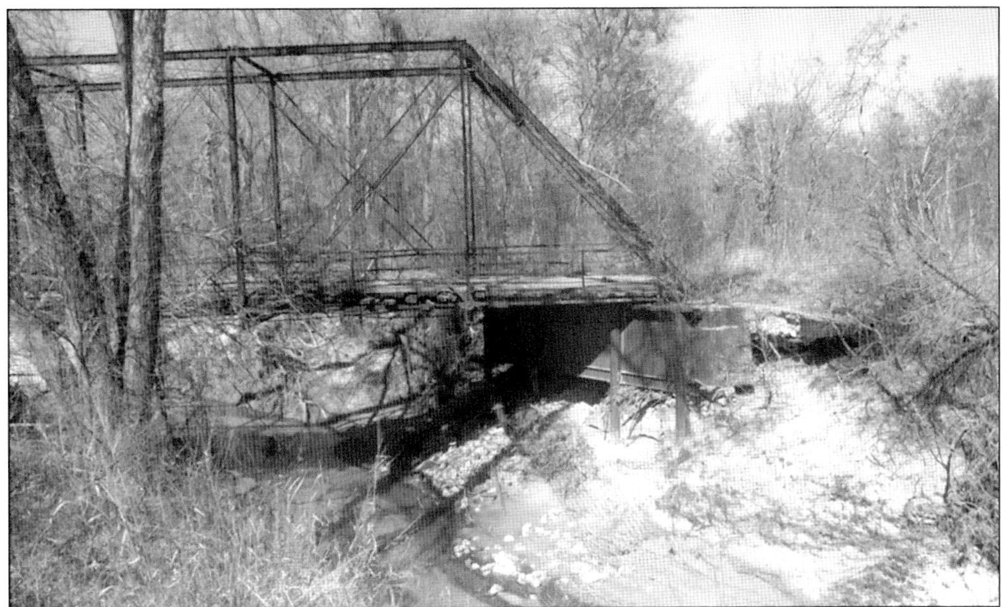

Hog Creek Bridge is a Pratt through truss bridge approximately 120 feet in length. Built during the latter part of the 19th century, this is one of several bridges that have deteriorated. Their presence reminds people of the critical role played by bridges in history. (Courtesy of David Galbreath.)

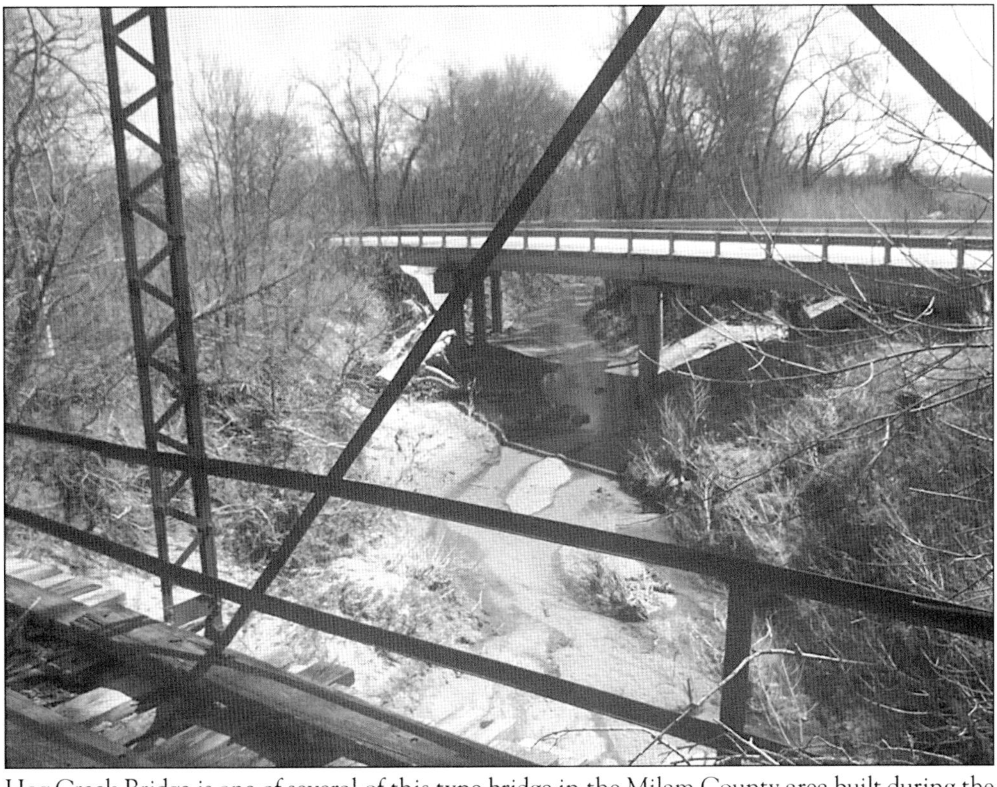

Hog Creek Bridge is one of several of this type bridge in the Milam County area built during the period from 1880 to 1920. (Courtesy of David Galbreath.)

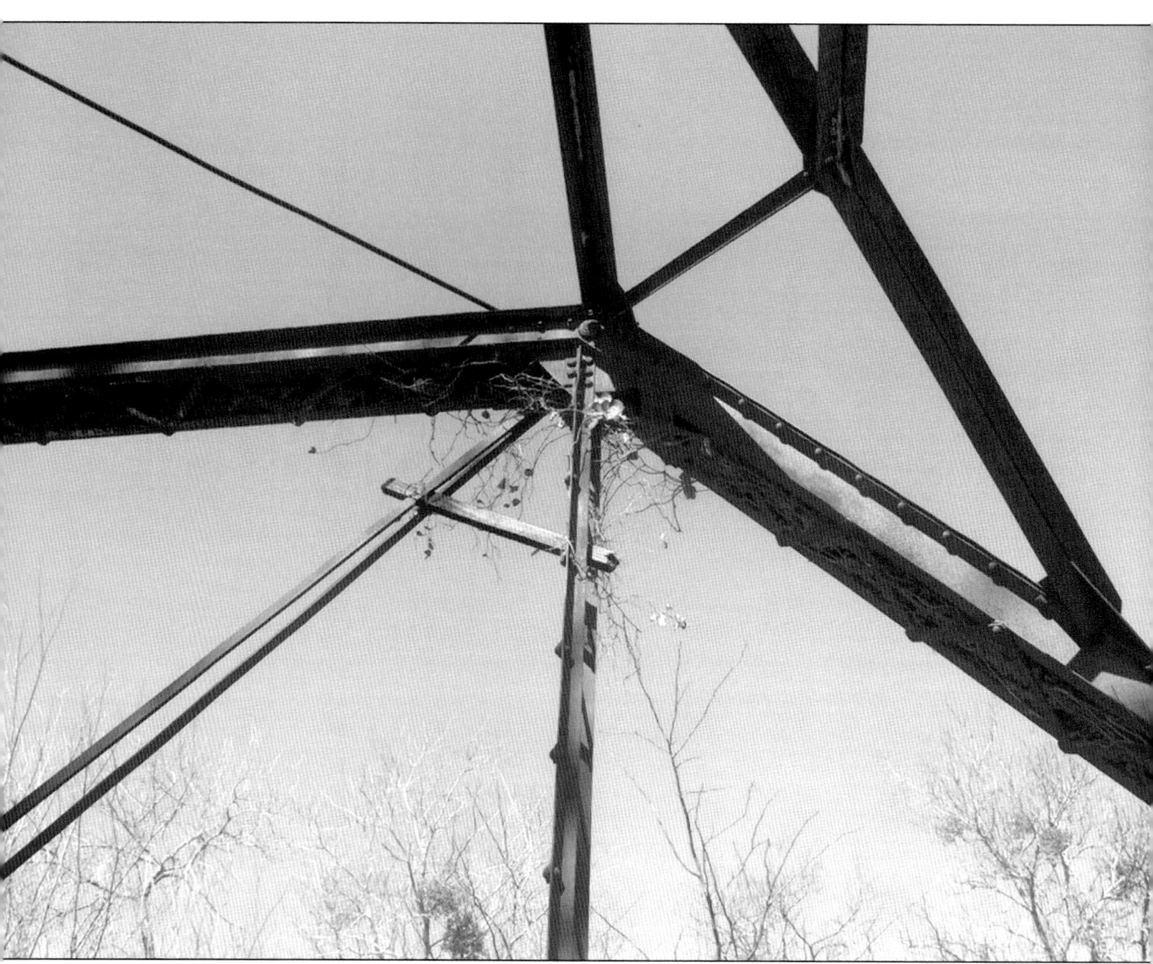

C.Q. Horton and the Chicago Bridge Company were the ones who most likely built the bridge in 1910. Although this is one of the lesser bridges in Milam County, it still carried great importance in facilitating transportation in the area. (Courtesy of David Galbreath.)

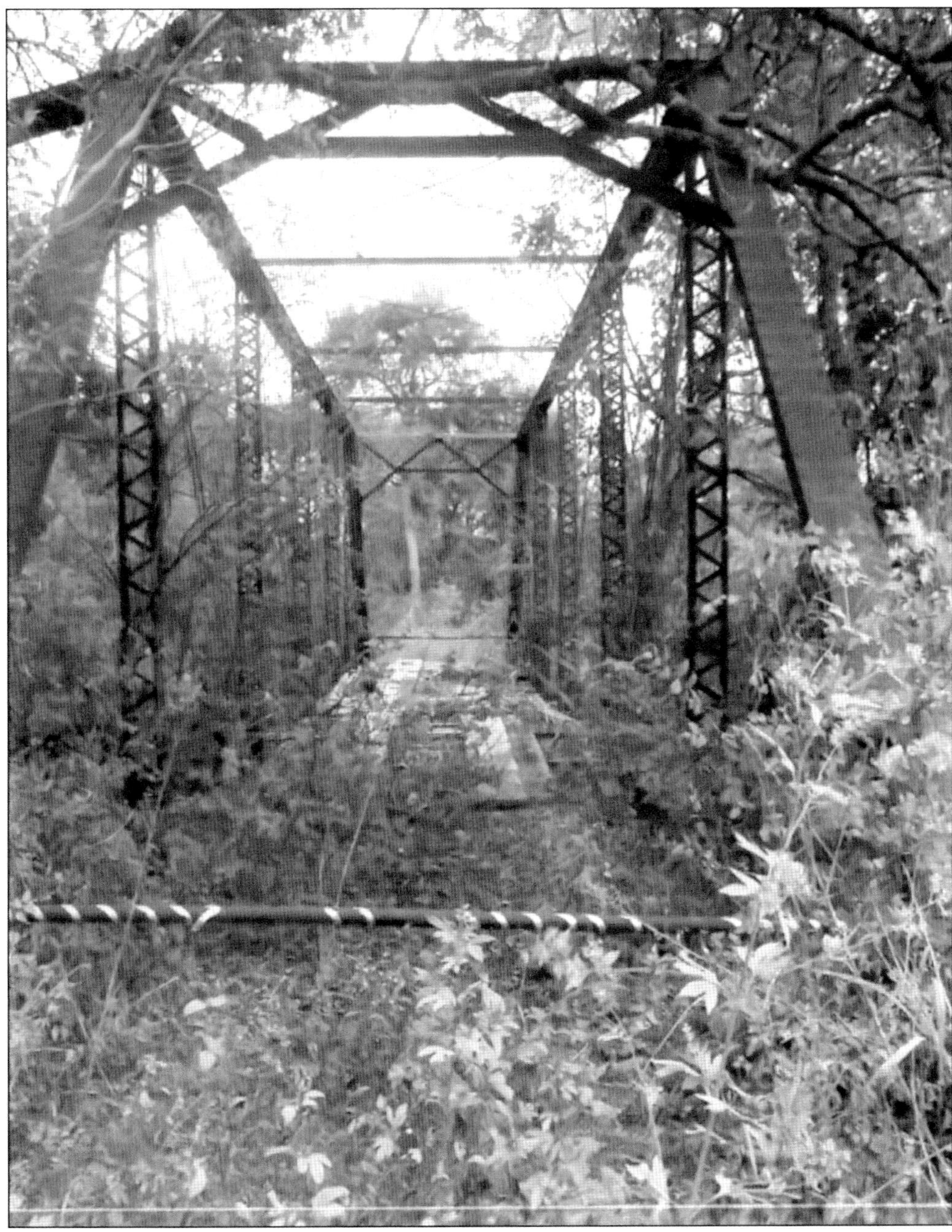

Hog Creek Bridge still stands but was bypassed with a cement slab bridge in 2004. The new bridge was erected off Farm to Market Road 2027. This type of bridge was built because of the relative ease of construction and its simplicity. (Courtesy of David Galbreath.)

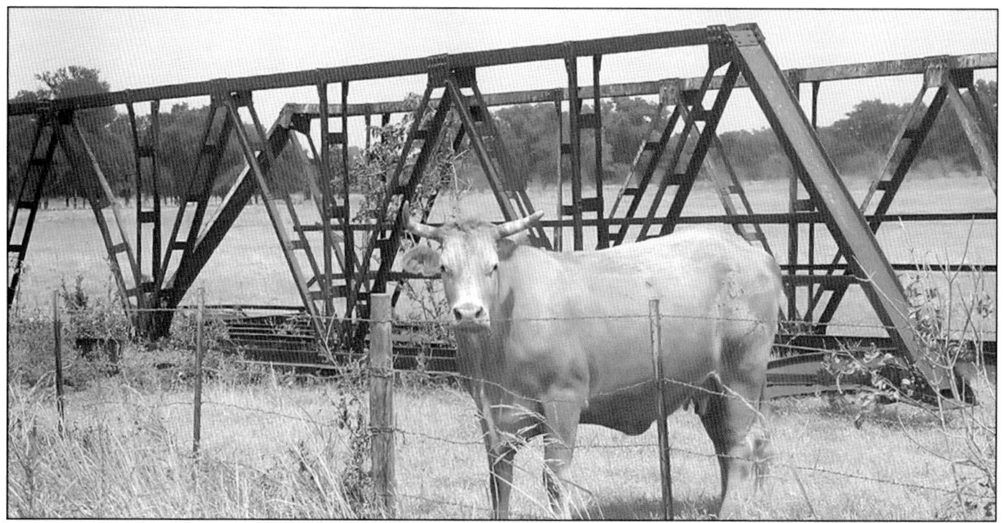

The above photograph, taken in 2015, shows Sheckles Bridge sitting in a pasture, as it has since it was replaced with a concrete bridge in 2003. In 1896, Ghent Sheckles petitioned the court for a bridge to be built over the San Gabriel River. The big flood of 1921 completely washed Sheckles Bridge away. In February 1923, at special meeting of the Commissioner's Court of Milam County, the court received the bridge over San Gabriel, known as Sheckles Bridge. Seen in the below photograph, Sheckles Bridge was located on County Road 429 across the San Gabriel River near Thorndale. This photograph shows Sheckles Bridge was in need of repairs. (Both, courtesy of Mike Brown.)

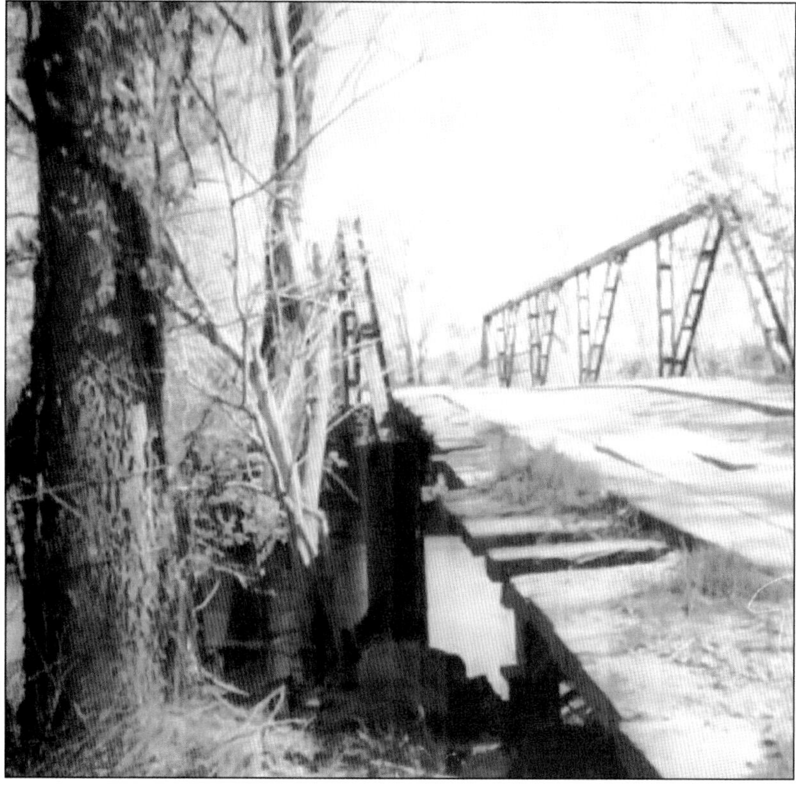

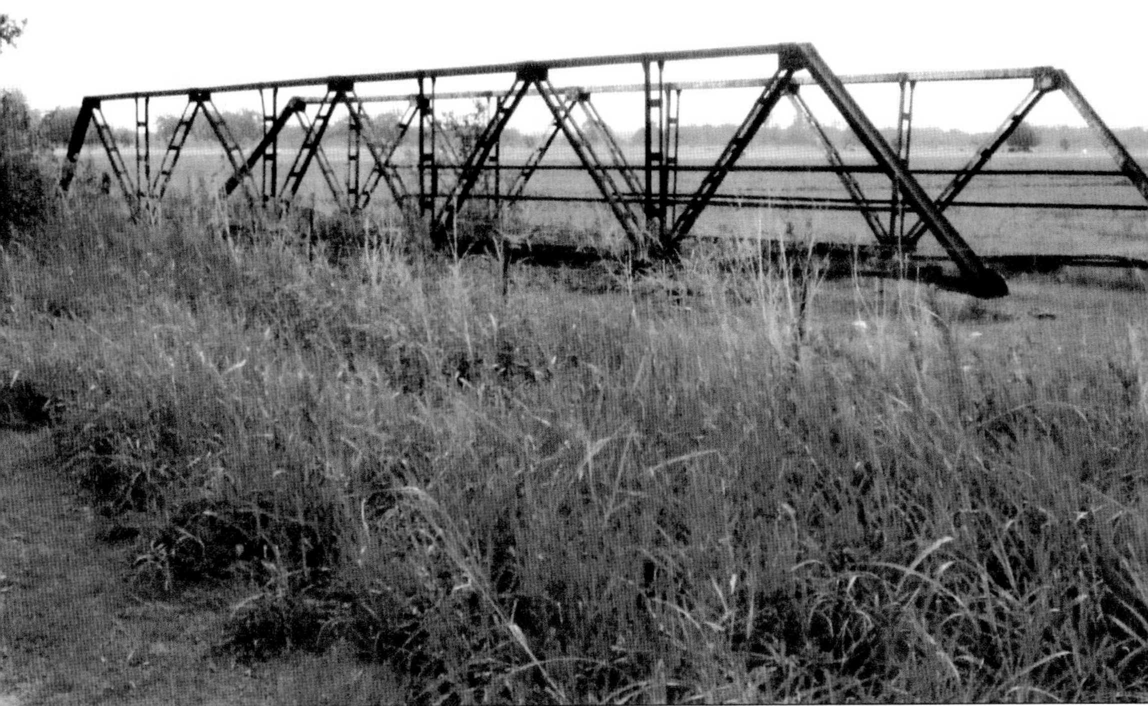

When Sheckles Bridge was beyond repair, it was removed and replaced with a concrete bridge in 2003. The old historic bridge was placed in a field off County Road 429 for about 12 years. (Courtesy of Jack Brooks.)

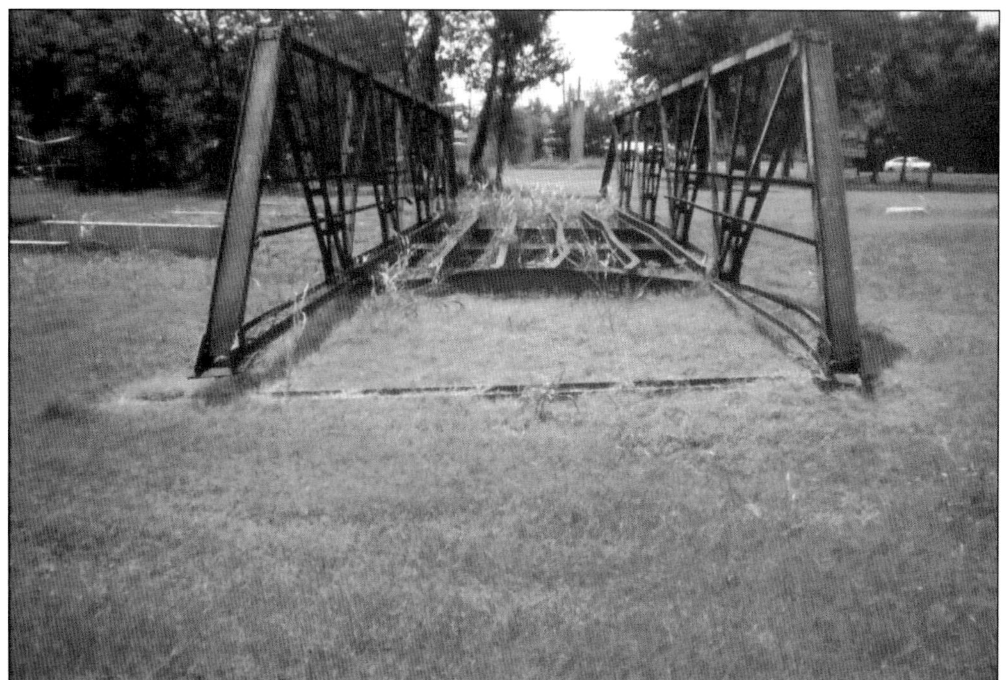

In 2015, Sheckles Bridge was relocated to the site of a new park being constructed in Rockdale. The city owns the land on the south side of US Route 79 just west of the Southwest Milam Water Supply office. (Courtesy of Joy Graham.)

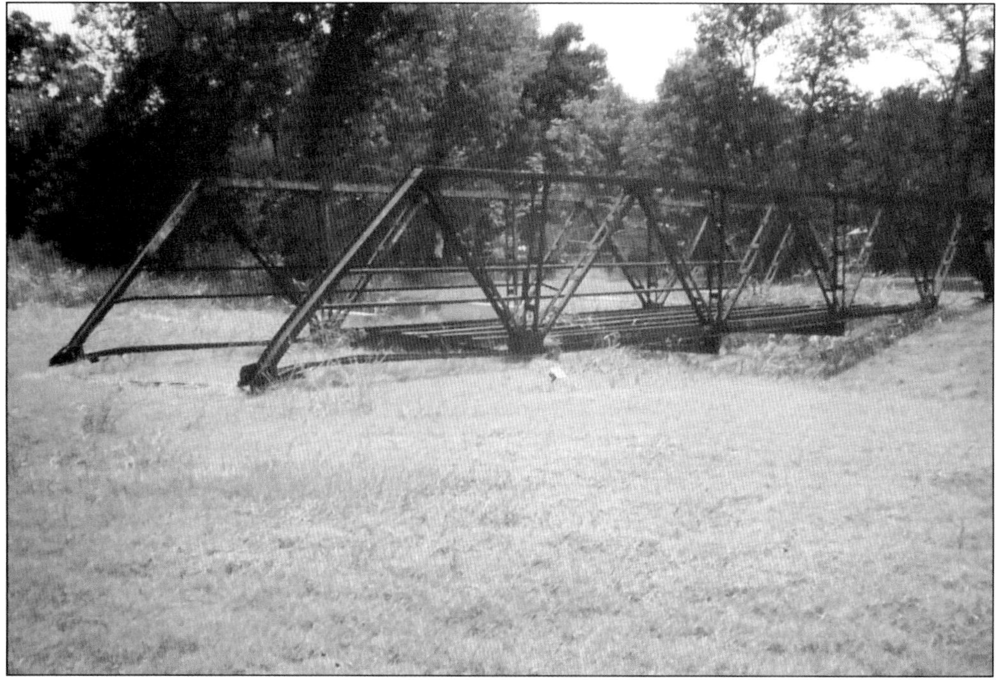

Sheckles Bridge is a Pony truss bridge that is 134.8 feet in total length, with deck width of 13.8 feet. It was constructed by J.F. Brown. The cost of building the bridge in 1923 was $2,146.70. (Courtesy of Joy Graham.)

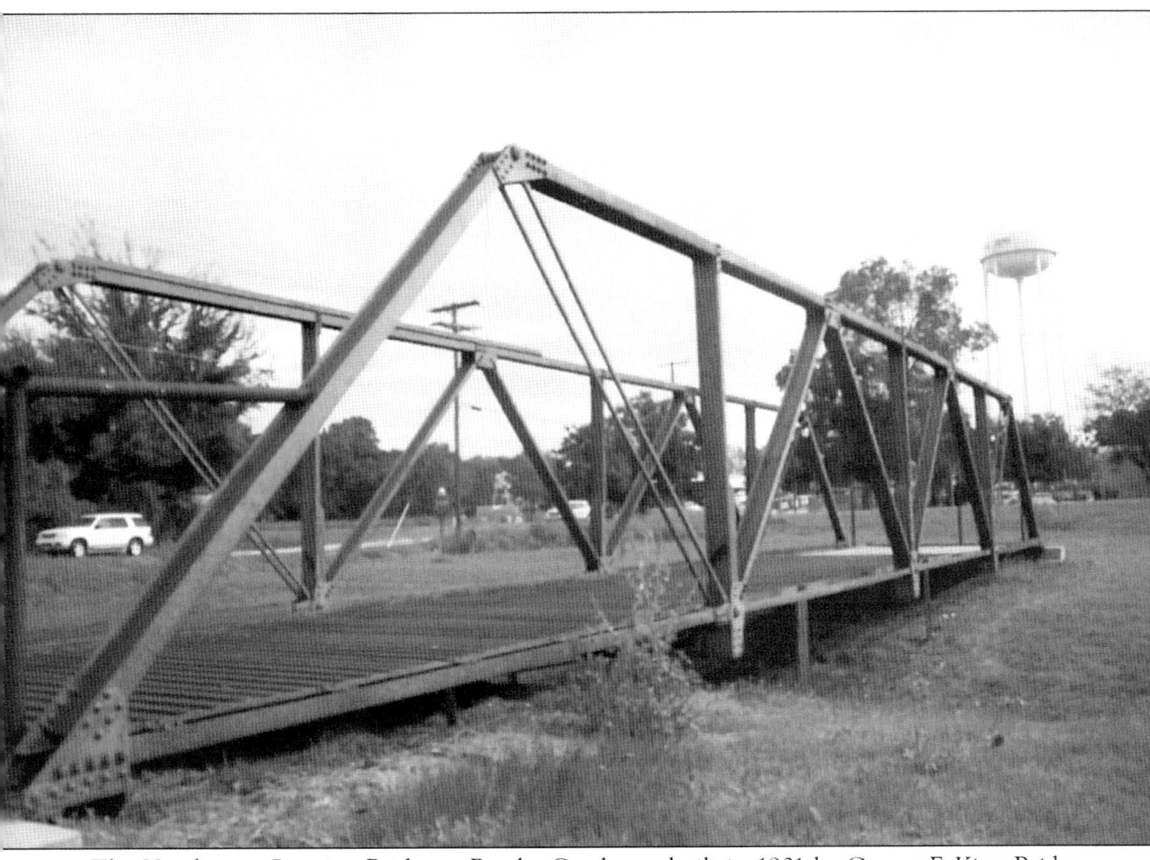

The Henderson Crossing Bridge at Brushy Creek was built in 1901 by George F. King Bridge Company. Rockdale citizens guaranteed the payment of $500 immediately upon completion. This historic bridge was the prime route to Georgetown on Henderson's Crossing between Rockdale and Thorndale on County Road 434. The bridge was the first built at this crossing near the community of Gay Hill. (Courtesy of Joy Graham.)

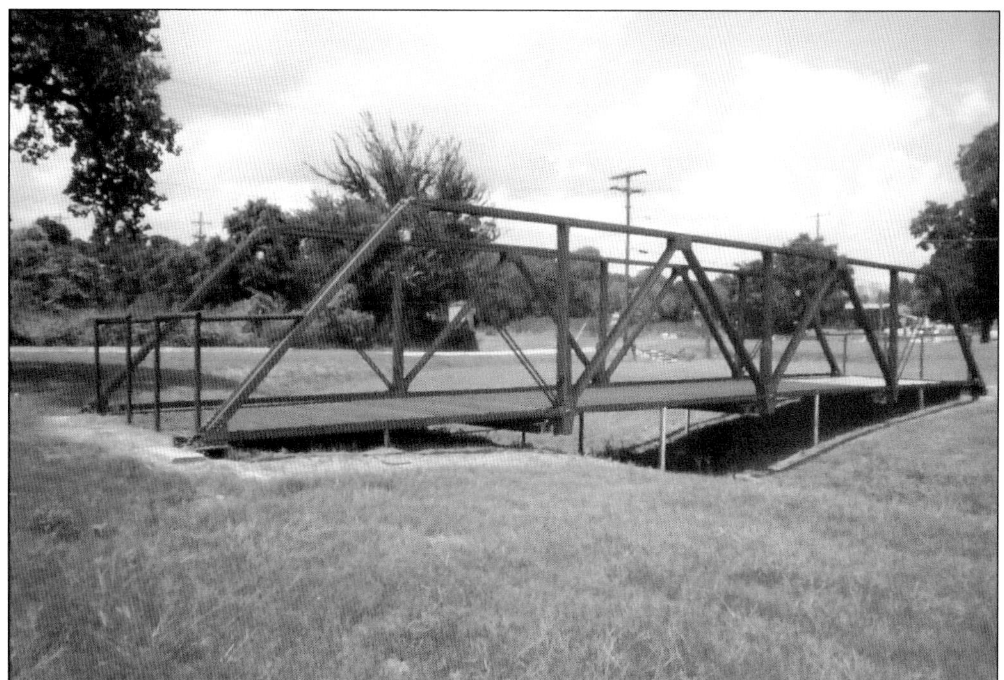

The Henderson Crossing Bridge is one of 29 pin-connected Warren Pony trusses surviving in the state. The bridge has a span of 94 feet in length with a deck width of 13 feet 2 inches. (Courtesy of Joy Graham.)

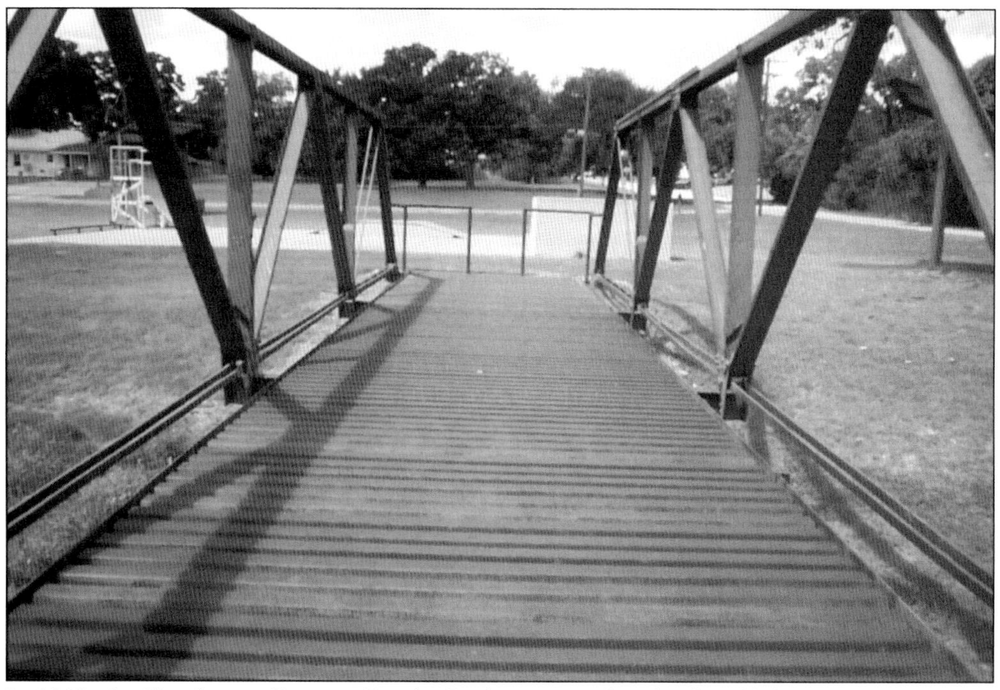

In 2002, the Henderson Crossing Brushy Bridge was replaced. The old, historical bridge was moved to Rockdale Skate Park as a pedestrian walkway at the corner of Mill and Wilcox Streets. A concrete slab bridge replaced the old Henderson Crossing Bridge. (Courtesy of Joy Graham.)

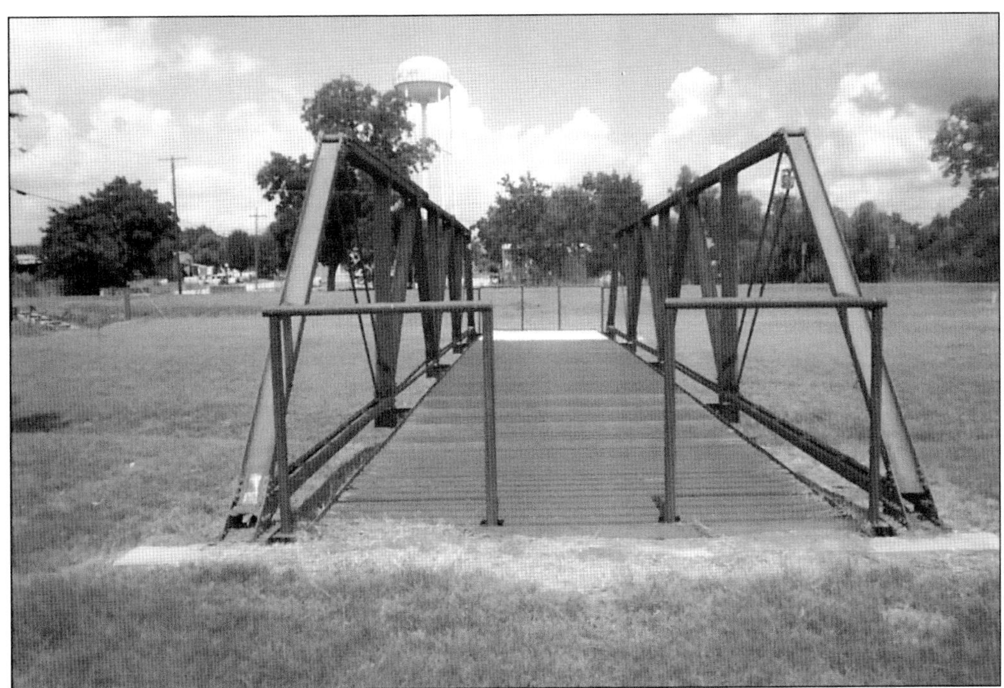

This photograph shows the entrance to Henderson Crossing Brushy Creek Bridge at Rockdale Skate Park. It has been modified to be used only as a pedestrian walkway. This bridge was eligible for the National Register until it was removed. (Courtesy of Joy Graham.)

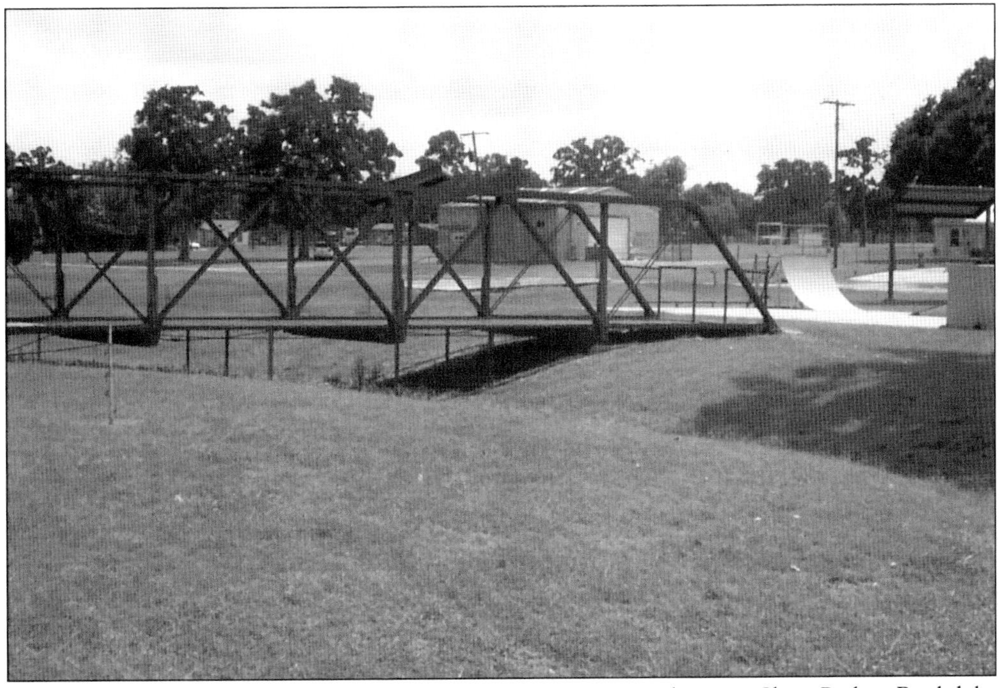

Henderson Crossing Brushy Creek Bridge made its way to its new home at Skate Park in Rockdale. In the background is a portion of Skate Park, where people gather to enhance their skateboarding skills. Now, this historic bridge must be supported by poles. (Courtesy of Joy Graham.)

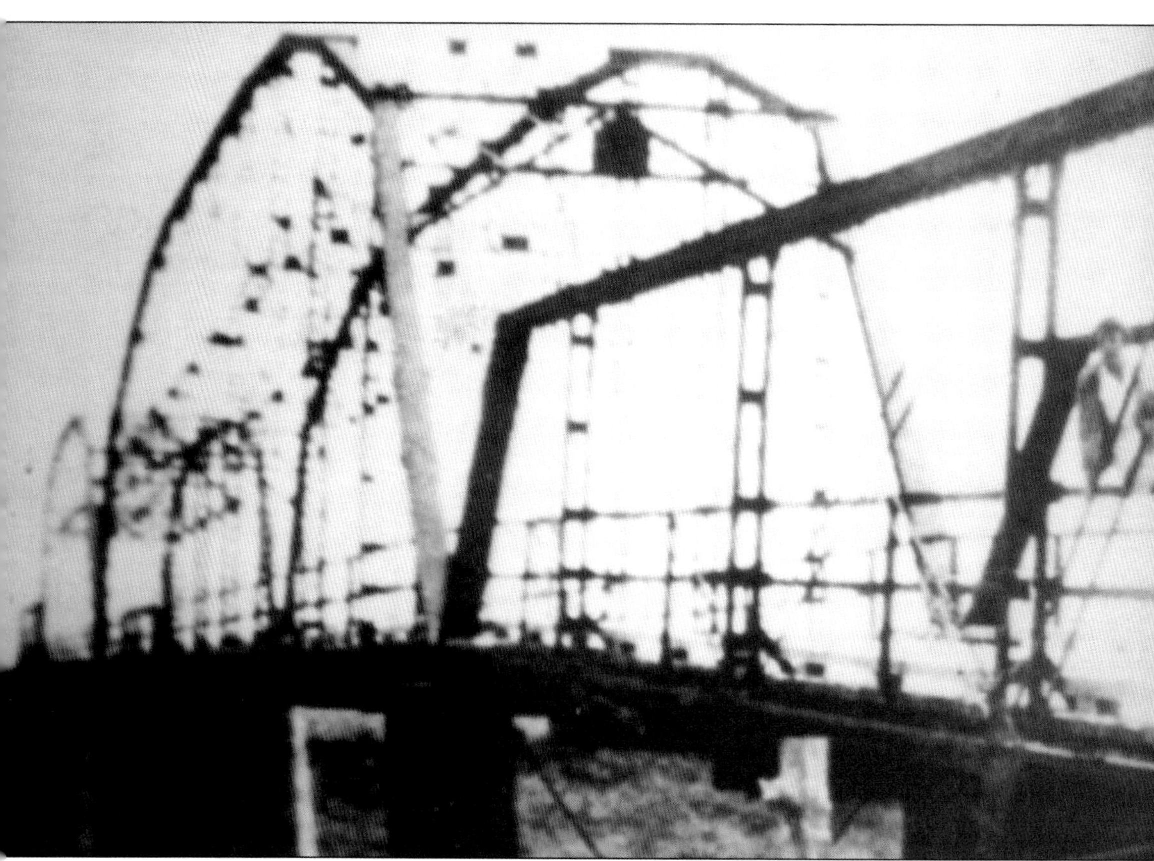

McCown Bridge was erected over a ferry crossing. Joshua Wilson McCown Jr. owned and operated a ferry crossing near the present Little River from 1876 to 1882. In May 1882, McCown petitioned the court to build a bridge at or near his ferry crossing. The story of the new bridge's construction began in 1883. The court contracted for the McCown Bridge with the Wrought Iron Bridge Company of Canton, Ohio. The bridge was to be 250 feet long with two spans of 100 and 150 feet, with approaches 14 feet in width. The bridge was to be completed on January 1884 at a cost of $17,100. (Courtesy of the Milam County Historical Museum.)

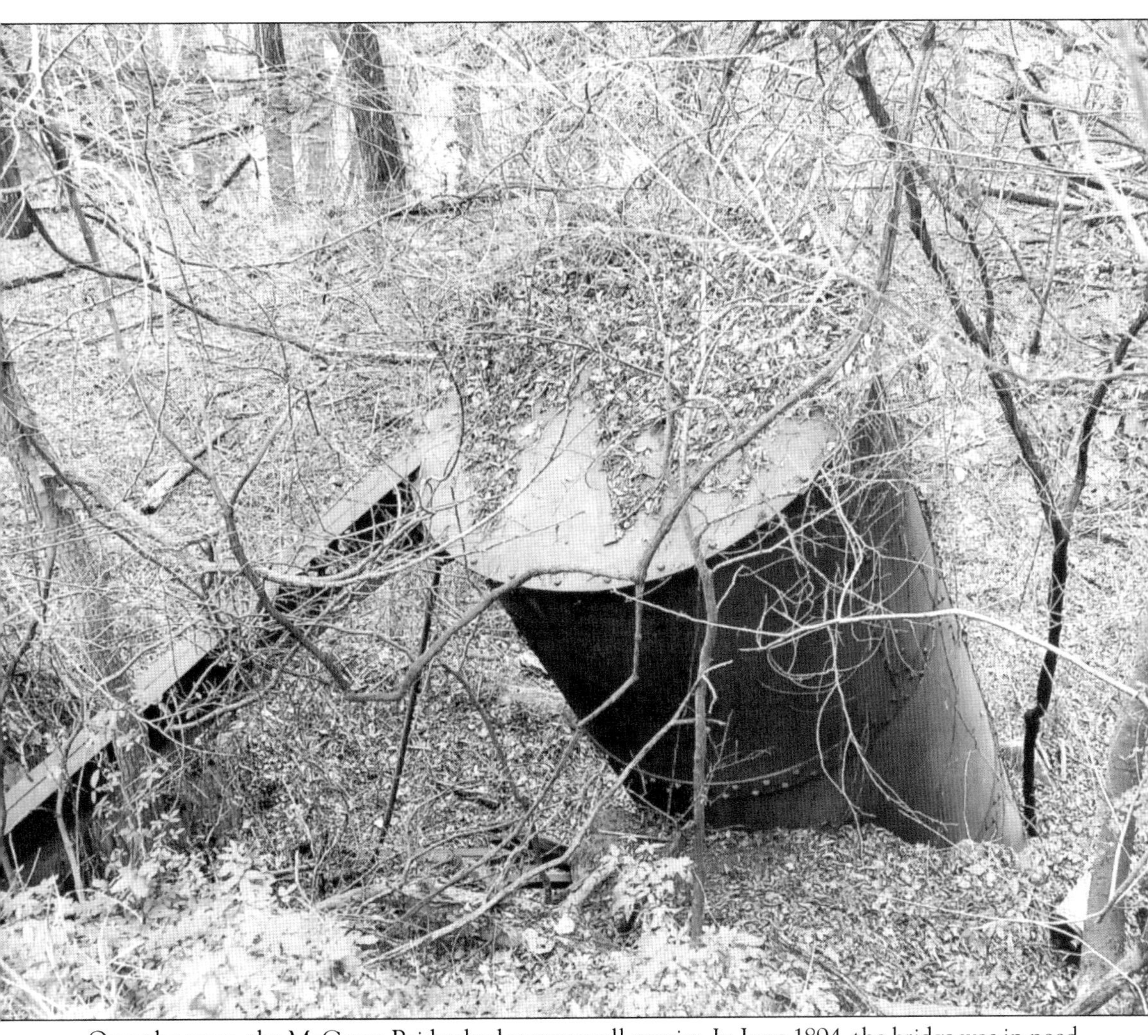

Over the years, the McCown Bridge had many small repairs. In June 1894, the bridge was in need of repairs where the safety of the traveling public and the preservation of the bridge were deemed as necessary. It was ordered that the wooden piers be removed and replaced with iron piers. The big flood of 1921 badly damaged the McCown Bridge, but it was still usable. No more repairs were made to the bridge, and it eventually fell into the river. The only evidence of its existence is a small portion of the iron piers, as shown in this photograph taken in 2015. (Courtesy of Jack Brooks.)

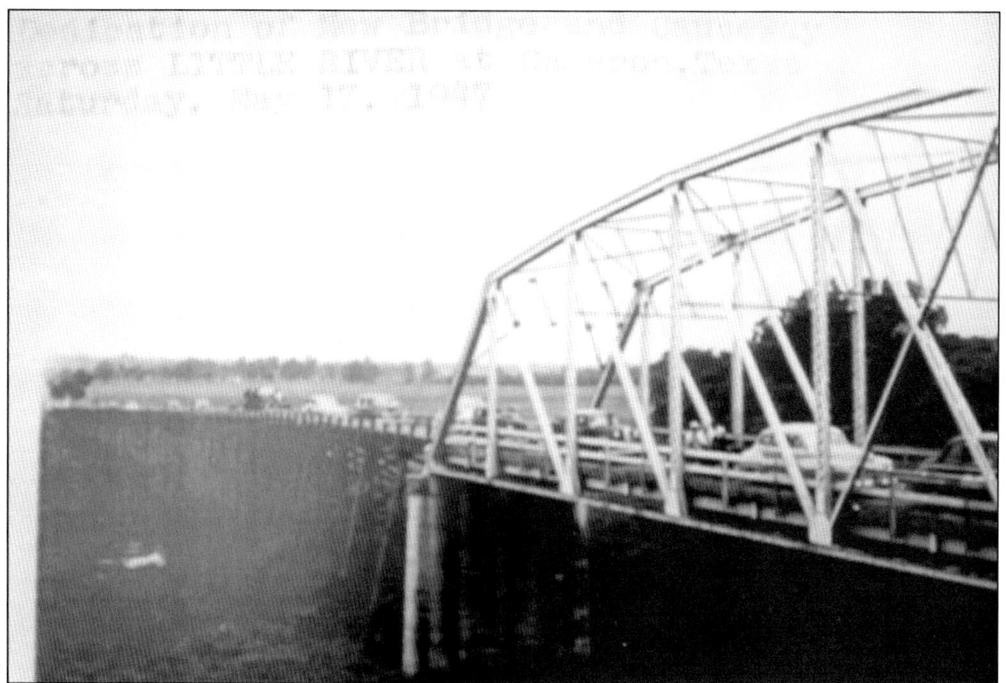

On February 11, 1943, the court issued an order to purchase private land for widening US Highway 77 and State Highway 36, enabling a major route through Milam County. After World War II, in 1947, close to the old McCown Bridge, the new mile-long Little River Bridge was dedicated. The photograph was taken from the old Little River Bridge on May 17, 1947. (Courtesy of the Milam County Historical Museum.)

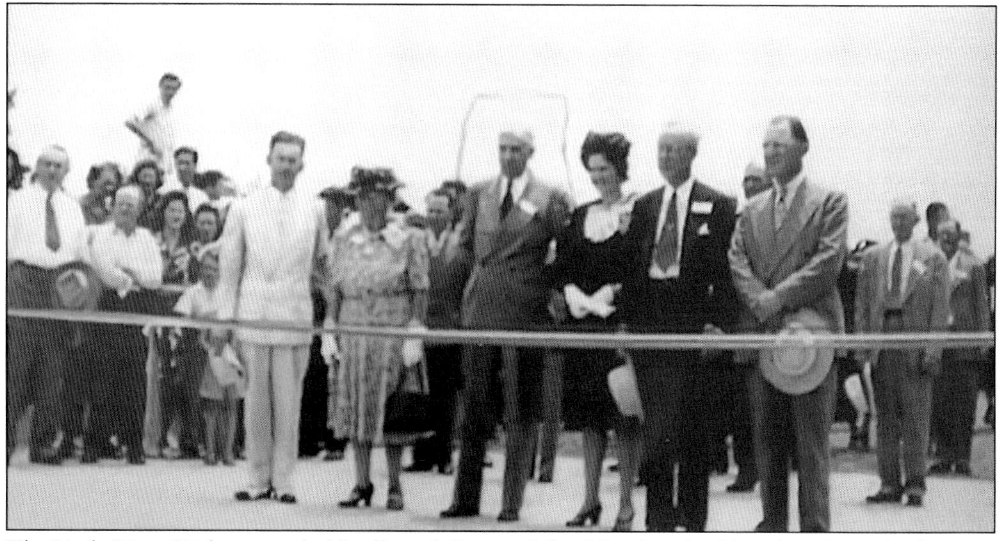

The Little River Bridge cost a half million dollars in federal funds to build in the 1940s. At the time of its dedication, on May 17, 1947, it was the fourth-longest bridge in Texas. When the bridge was contracted, Jeff T. Kemp was the Milam County judge. He died November 3, 1946, but his widow, Lina Rodgers Kemp, attended the dedication and cut the ribbon The people standing in front of the crowd are, from left to right, unidentified, Lina Kemp, Judge Dan Tyson, two unidentified, and Sen. Kyle Vick. (Courtesy of the Milam County Historical Museum.)

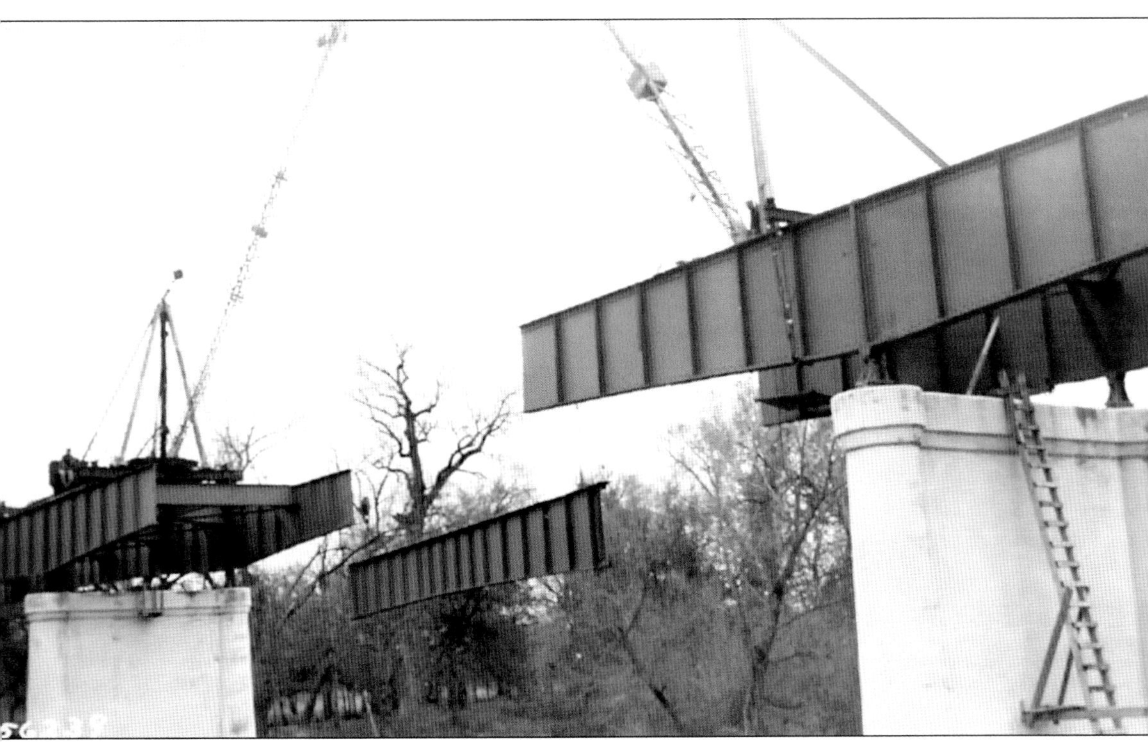

This photograph was taken December 10, 1946, during the construction of the Little River Bridge. They are raising the upstream center section 330 feet. In case there had been some mistakes in the calculations concerning the behavior of the foundation or the pressure expected from floods, the cement piers were set three years earlier. However, there were no errors—just another tribute to the skill of highway engineers whose plans were not upset by these forces of nature. (Photograph by Victor R. Schmidt, courtesy of George and Eleanor Schmidt Hollas.)

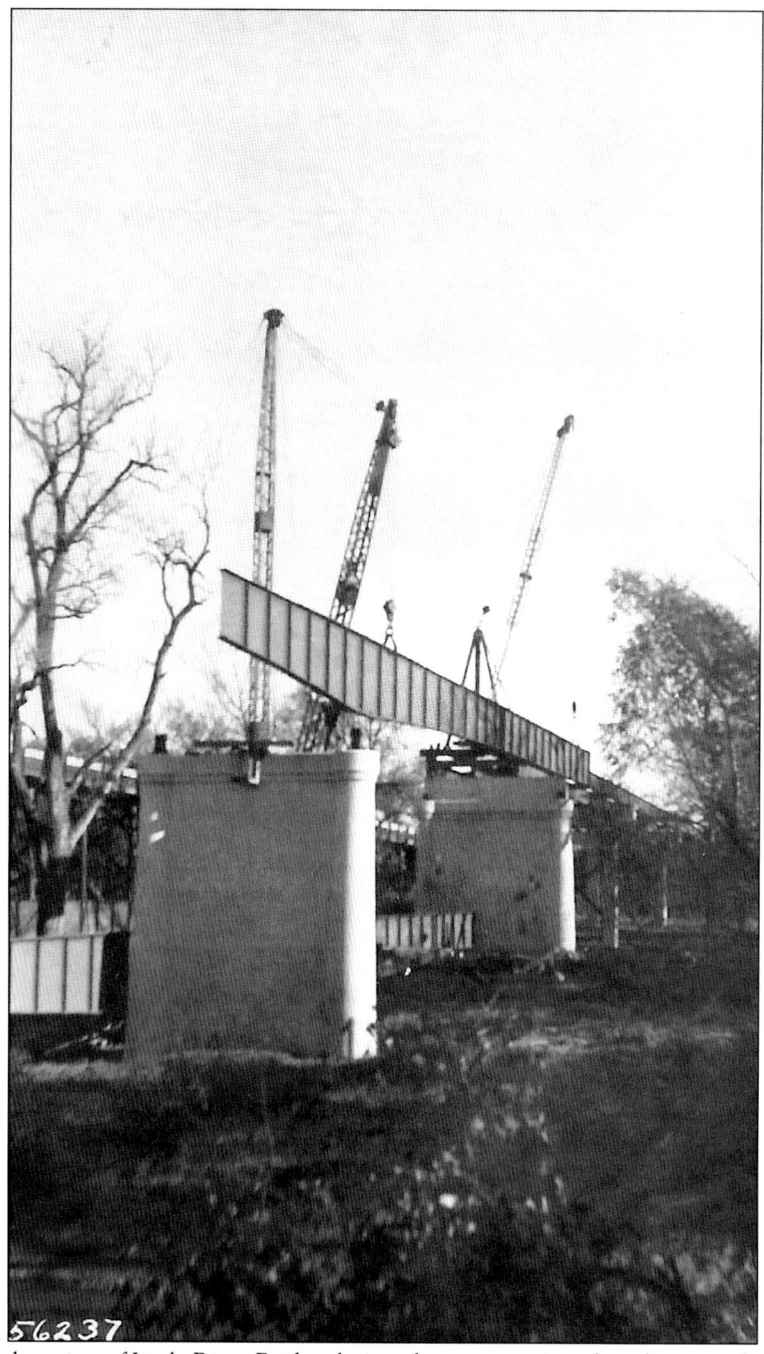

This is another view of Little River Bridge during the construction that shows workers setting a double section of 330-foot plate girder between piers No. 64 and No. 65. It is another tribute to the engineering skill of those who built the piers that no difficulty was encountered in fitting the steel spans to the piers. This is in spite of the fact that the piers had been exposed to the forces of the elements for more than three years, during which a settlement or shifting of a fraction of an inch would have complicated the erection of the steel spans. (Photograph by Victor R. Schmidt, courtesy of George and Eleanor Schmidt Hollas.)

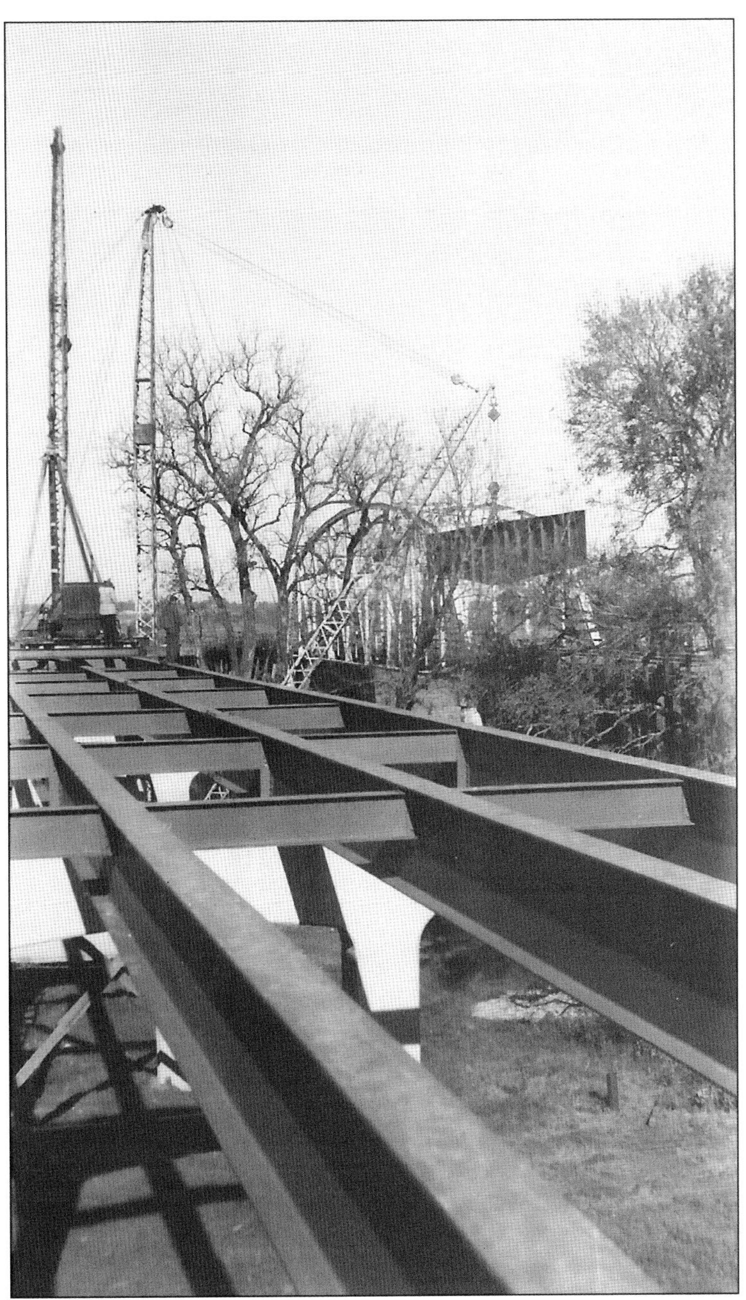

Little River Bridge was built over the Little River because the bottomland was notoriously flood-prone east of Cameron. The bridge was 10 years from studies to completion, delayed in 1943 because steel could not be obtained during World War II. In 1943, there was plenty of sand, gravel, and concrete, so the piers of the bridge were built in order that they would be ready to carry the spans the minute steel was available. Little River Bridge measurements remained true so that the ends of the steel spans rested on the center of the plates designed into the piers at the time there were built in 1943. The crane is raising a section of a 300-foot plate girder unit off the truck on the old bridge on November 28, 1946. (Photograph by Victor R. Schmidt; courtesy of George and Eleanor Schmidt Hollas.)

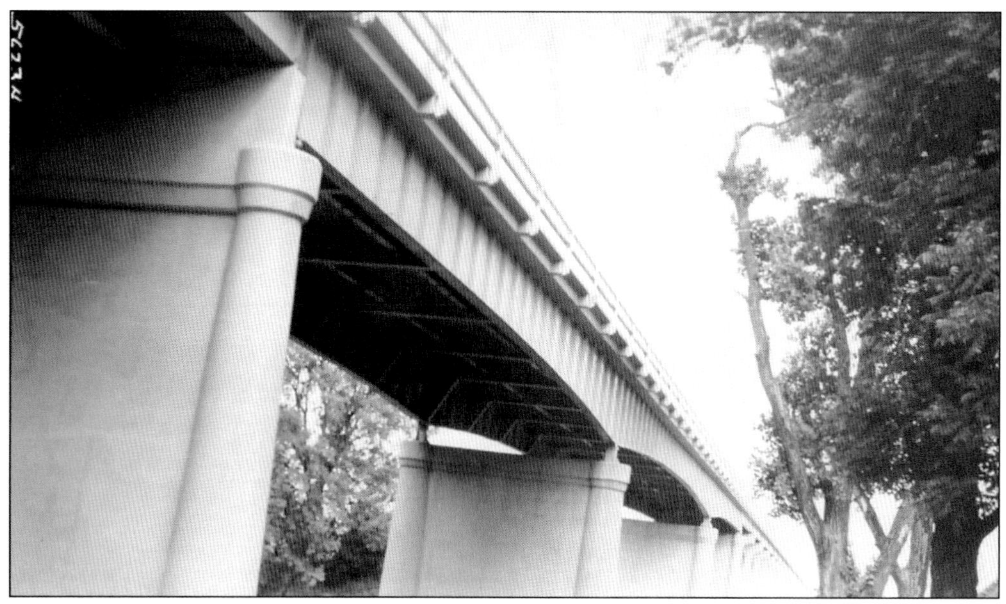

This view looking up at concrete piers of Little River Bridge shows the underside of the bridge. The construction of the piers was performed by the Southern Contracting Company of Austin in 1943. Victor R. Schmidt was senior resident engineer with the state highway department that supervised the construction by Thomas & Ratliff of Rogers. It began in December 1945 and was completed in late April 1947. The steel was fabricated in the Dallas Plant of Mosher Steel Company. (Both, photograph by Victor R. Schmidt, courtesy of George and Eleanor Schmidt Hollas.)

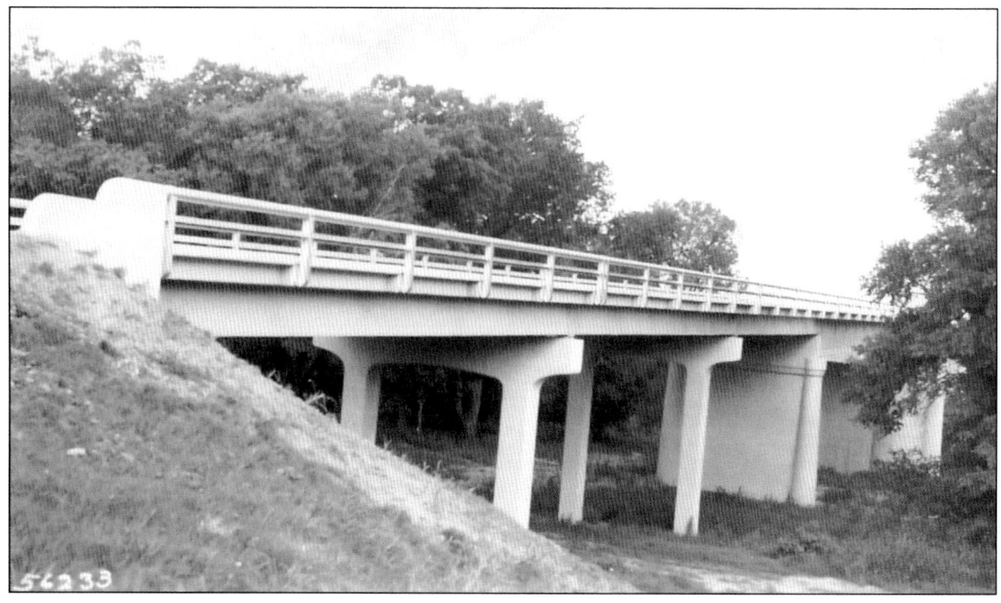

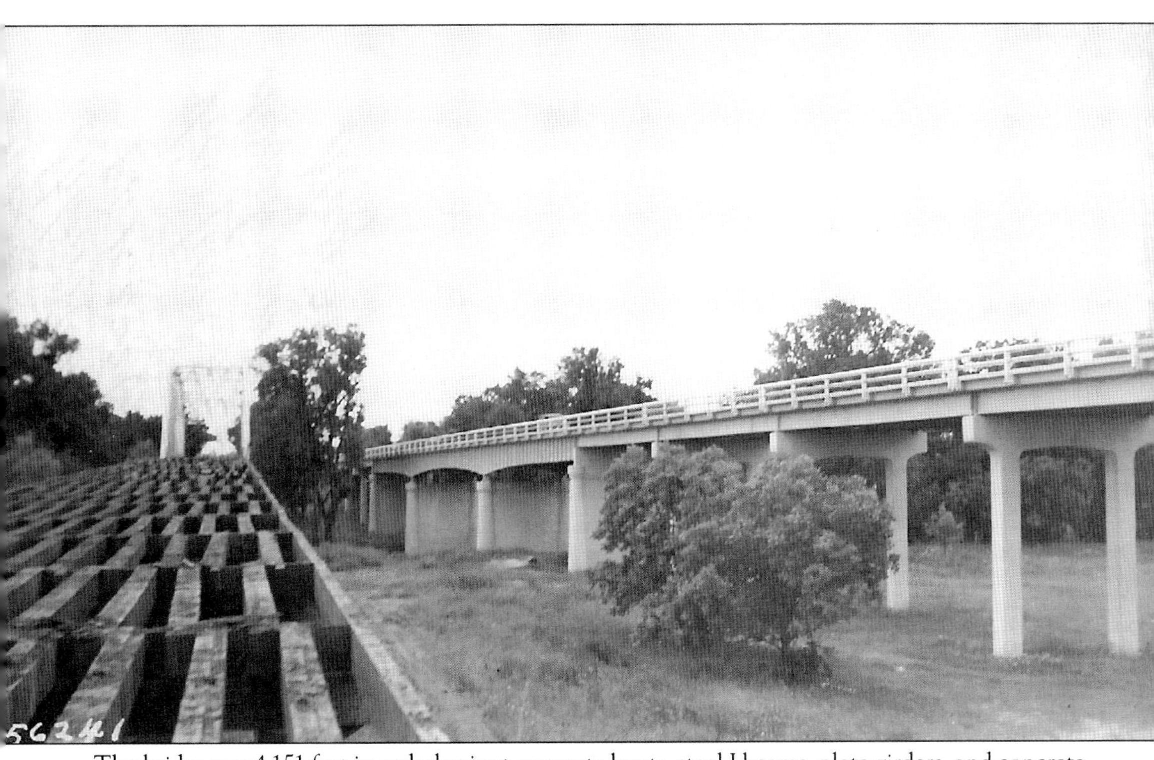

The bridge was 4,151 feet in orderly giant concrete bents, steel I beams, plate girders, and concrete decking almost a mile in length. It was constructed above all known floodwater levels to insure travel at all times over Highways 77 and 36 through Milam County. Through the years, the once smooth concrete became an uneven decking, with tires taking a beating. The old bridge was too narrow for two-way traffic, and maintenance of the concrete decking was costly. After 58 years, the bridge was imploded in 2005. The photograph shows the old Little River Bridge being torn down. (Photograph by Victor R. Schmidt, courtesy of George and Eleanor Schmidt Hollas.)

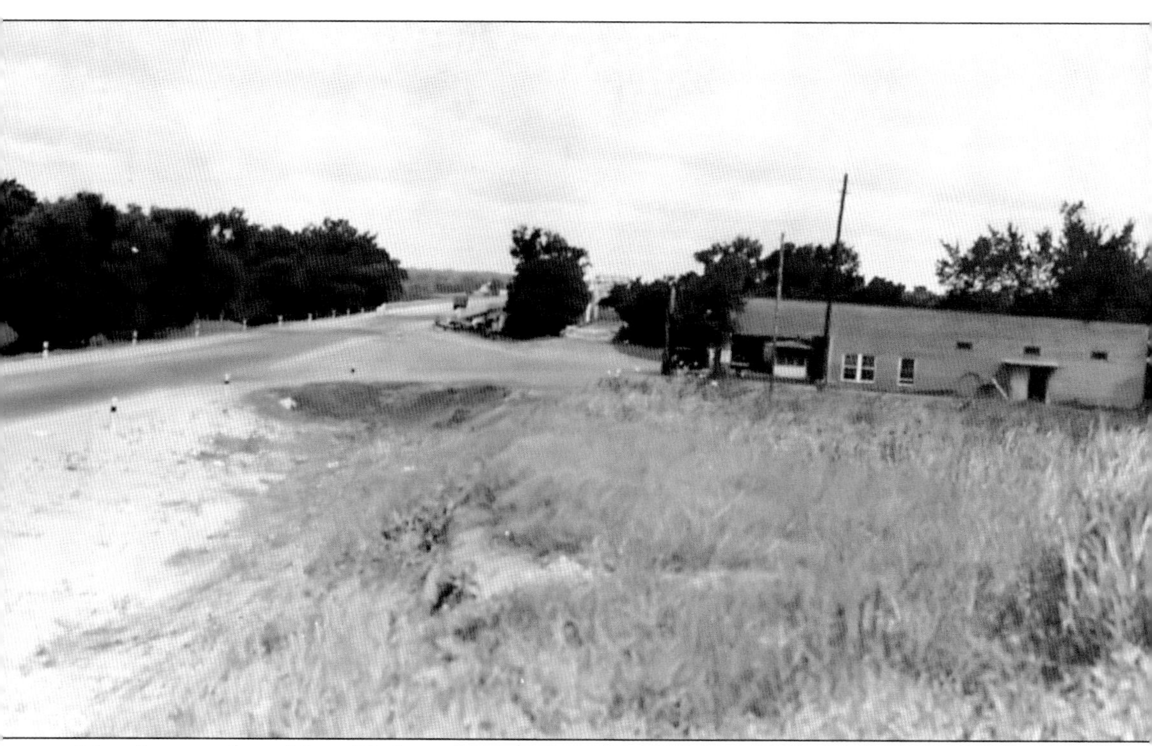

This photograph is of State Highway 36 and US Highway 77 crossing over the Little River Bridge. Lamkin's General Store is on the right side of the highway. This store was a popular rest stop for travelers who wanted to purchase canned food, general merchandise, fuel, and kerosene. The building is now vacant; however, its history will never be forgotten. (Courtesy of George and Eleanor Schmidt Hollas.)

Four

Other Significant Bridges

Chapter 4 focuses on other significant bridges in the area. Even though some of these are smaller bridges—and in some instances, no traces remain—these bridges were extremely significant to Milam County in their day. Without them, some areas of the county would have remained undeveloped and without access and certain communities would not have flourished. Among the bridges that depicted are San Andres, Cummings Creek, San Gabriel, Little River, Black Bridge, Holtzclaw, Hendrix Crossing, Clay Creek, Cedar Creek, and Walkers Creek.

Bridges had been authorized in the early days of Milam County. The county's location and landscape mandated that attention be given to bridges, hence permission was granted to organize a bridge company. The Calvert Bridge Company was formed and given license to raise money for a bridge that would connect Milam and Robertson Counties.

Building continued, and in 1871, James Holtzclaw was given the authority to build a bridge. Thus continued the orderly progression of transportation across Milam County. This chapter contains details of the advancement focusing primarily on smaller bridges that are no longer standing.

As the march of bridges concludes, think for a moment of their significance. They have seen Milam County grow from an agrarian wonderland into a significant source of coal, resulting in the presence for many decades of the Aluminum Company of America (Alcoa).

These were indeed bridges with a "view of history," a phrase coined at the June 2014 dedication of the restored treasure Worley Bridge. Cherish this idea, and recognize with the National Trust for Historic Preservation the history happened here—preserve it for generations yet unborn.

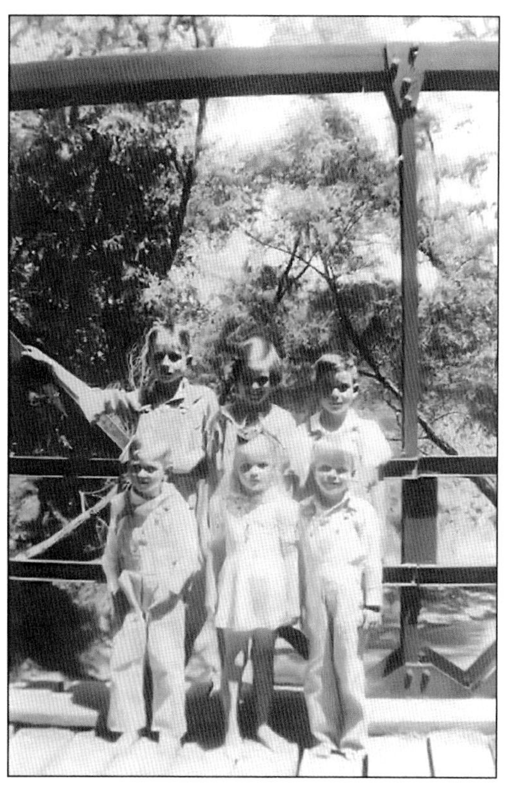

San Andres was a small town located eight miles northwest of Rockdale. It was laid out in the early 1850s, and its post office was established in 1852. The townspeople built a Methodist church, but it was destroyed in the flood of 1921. There are no physical remains of the town except for the San Andres Cemetery. Descendants of the original settlers still live in the area. Pictured here around 1938 are early San Andres residents, from left to right, (first row) Graham Young, Barbara Young (Kastner), and Ollie Young; (second row) George H. Young Jr., Bessie Jo Young (Farrar), and John Wallace Young. (Courtesy of Steve and Lynn Young.)

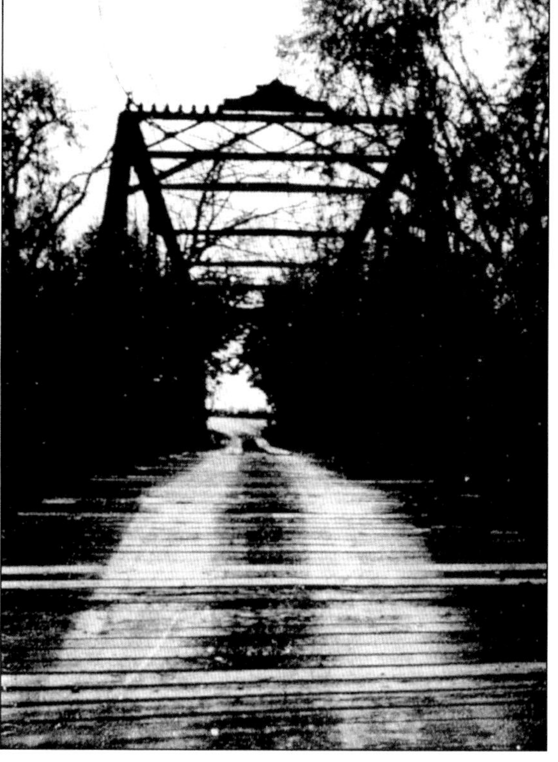

Cummings Crossing Bridge was built in 1891. It was a Pratt through truss bridge located off Farm to Market 1915. Technically, it is over Donahoe Creek at Cummings Crossing. Its location is also in the vicinity of Clay Creek. The bridge was named for the Cummings family who lived in that area, and family members still remember the fun they had during the summer as they enjoyed the water and the bridge. (Courtesy of the Milam County Historical Museum.)

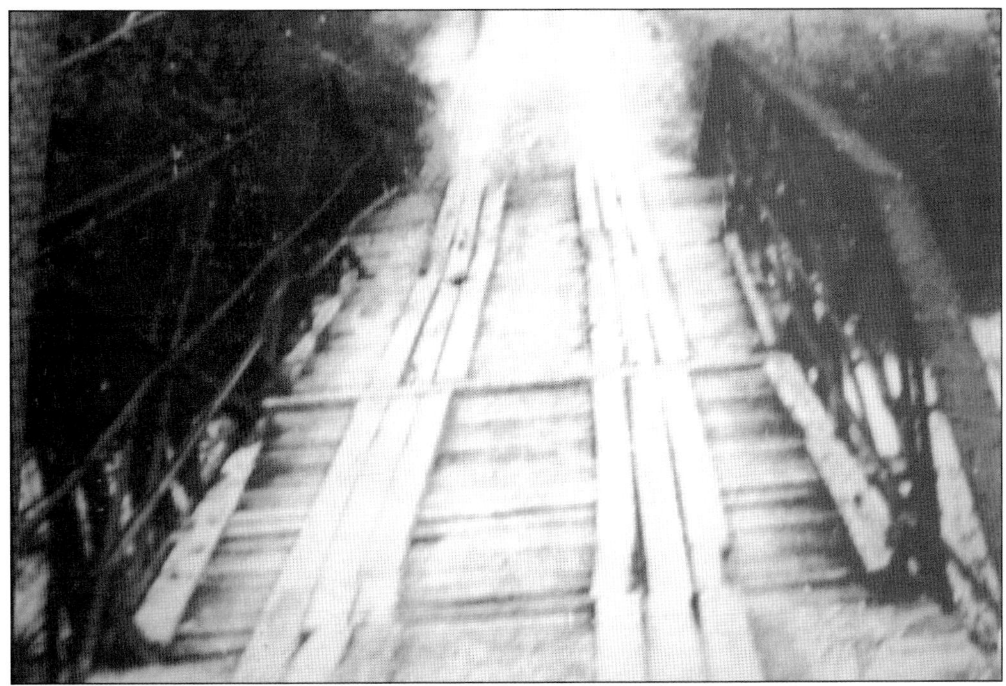

San Gabriel Bridge was constructed on Farm to Market 486, the road that connects Thorndale to San Gabriel. The bridge was built to give the area residents access to the village of San Gabriel. By 1880, the village had a church, two schools, and 130 residents. The bridge served the townspeople well for years. After an accident occurred on the bridge, it was completely rebuilt. (Courtesy of the Milam County Historical Museum.)

Sandy Creek Bridge is one of several bridges built to accommodate people on both sides of the San Gabriel River. In recent years, the bridge fell into disrepair and was the scene of a fatal accident in the late 20th century. After it was rebuilt, citizens continued to use the bridge on a daily basis. It is now maintained by the county. (Courtesy of the Milam County Historical Museum.)

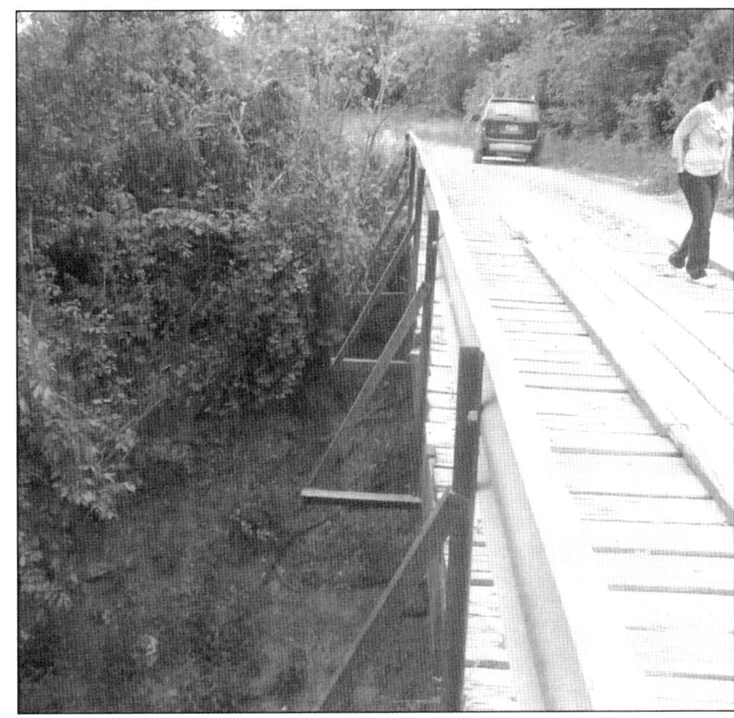

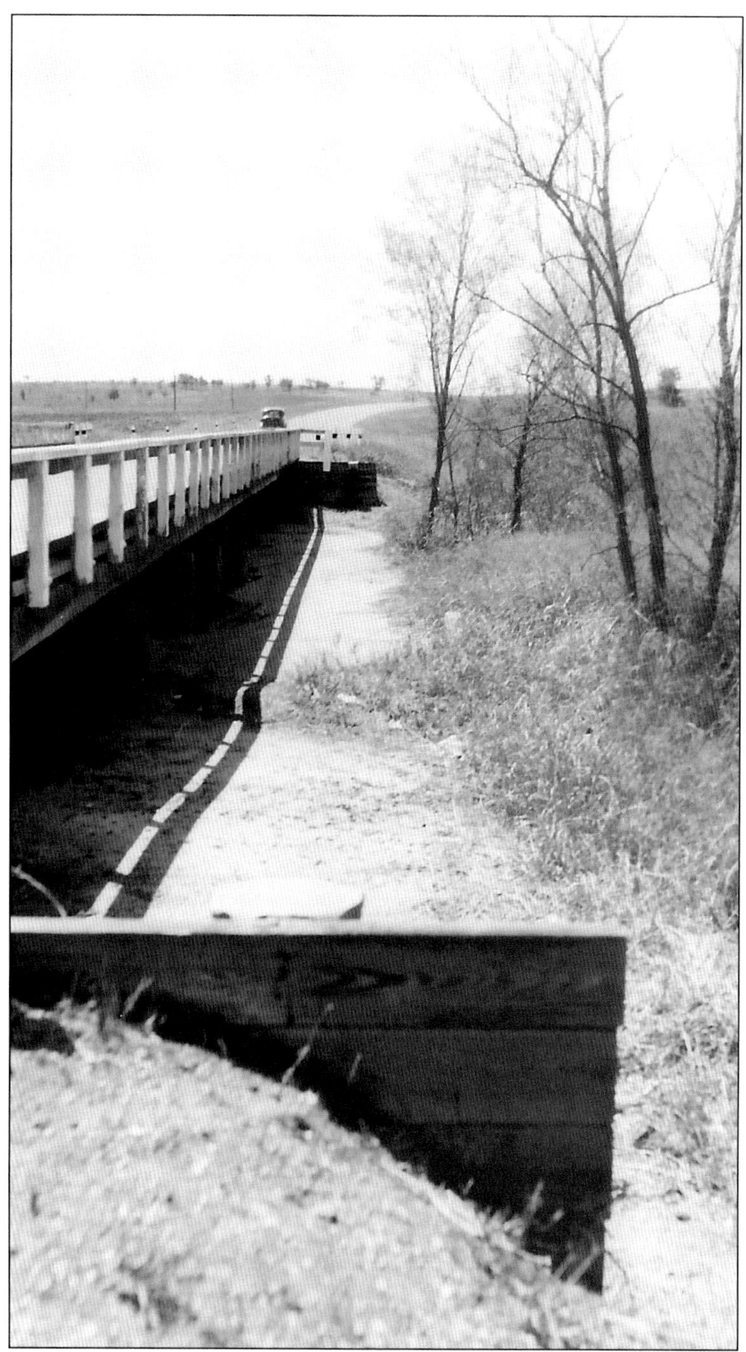

In 1884, the county built iron bridges across Walkers Creek. The creek begins a half mile southwest of Burlington, which is in eastern Milam County. The contract was given to the Wrought Iron Bridge Company to use earthen and wooden approaches. The bridge was named for W.H. Walker, who was a county judge and had land on the creek in 1836. In 1894, Milam County contracted with the New Columbus Bridge Factory for bridges to be built across Elm and Walkers Creeks. Walkers Creek was among several bridges financed by issuance of $19,800 in bridge bonds. (Photograph by Victor R. Schmidt, courtesy of George and Eleanor Schmidt Hollas.)

Black Bridge was located on the Brazos River off Farm to Market 979 and is near Crossroads and Calvert Junction. It is also close to the Calvert Oil Field. This was originally a wooden toll bridge built by August Clinton Black, grandfather of onetime county sheriff Carl Black. (Courtesy of Joy Graham.)

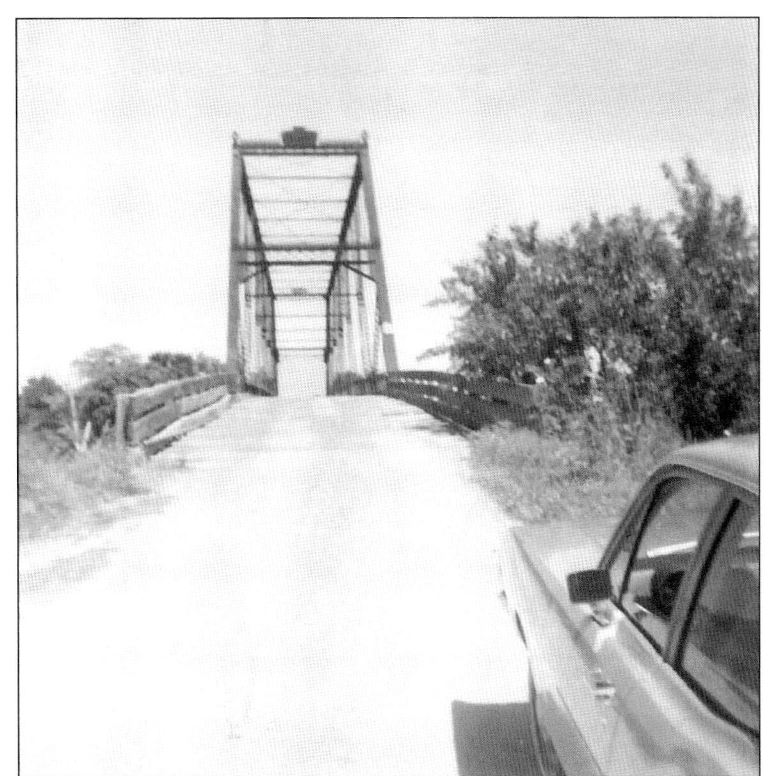

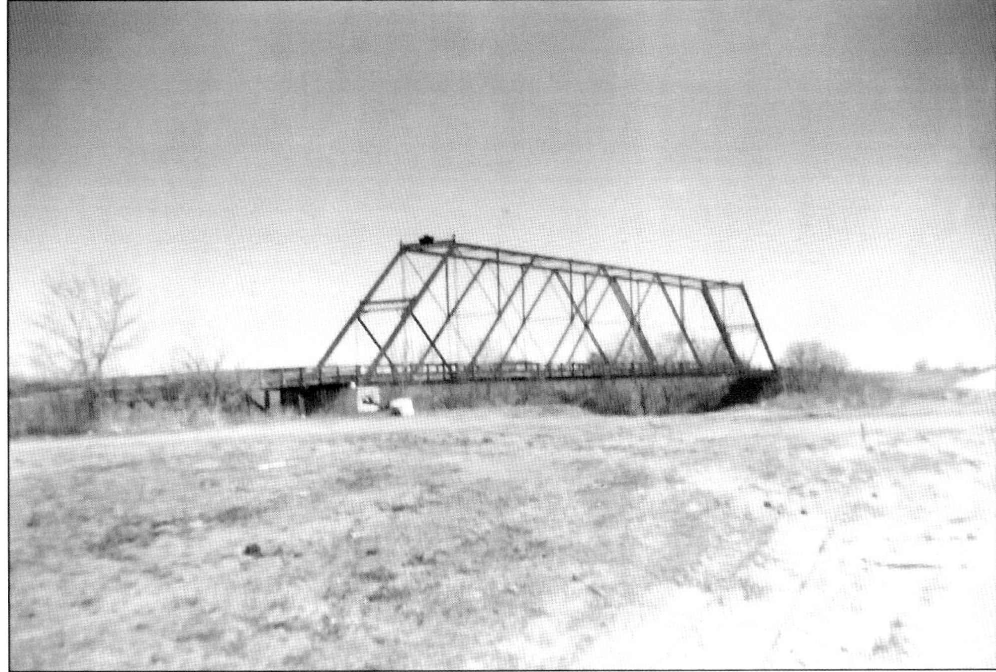

The Black Bridge Cemetery is located just past Black Bridge. The average elevation is 282 feet. This tall and elaborate bridge connected Milam County and Robertson County and was built in 1900. (Courtesy of George and Eleanor Schmidt Hollas.)

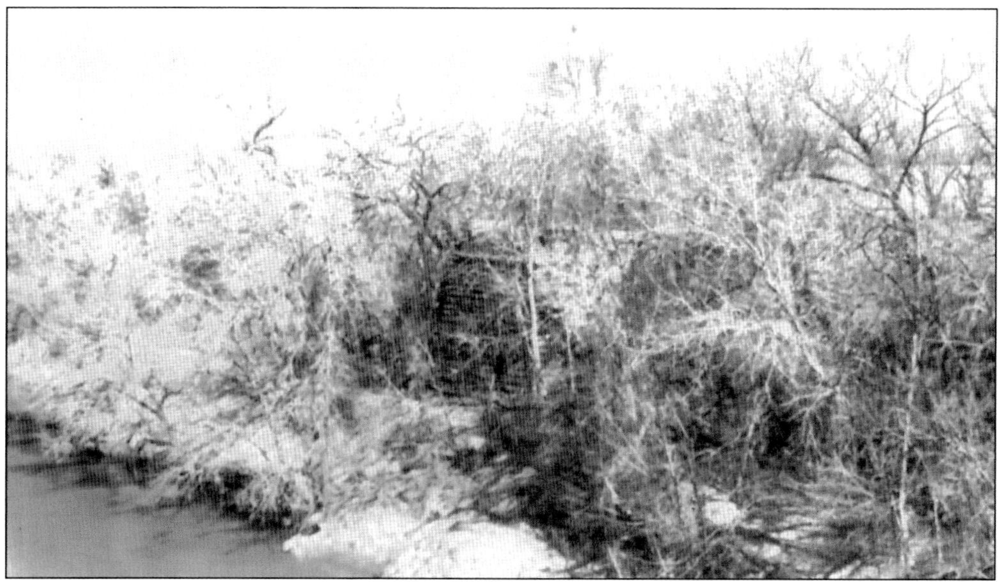

This photograph shows an old pier from Black Bridge on Farm to Market 979 at the Brazos River. The span was 370 feet long. (Courtesy of David Galbreath.)

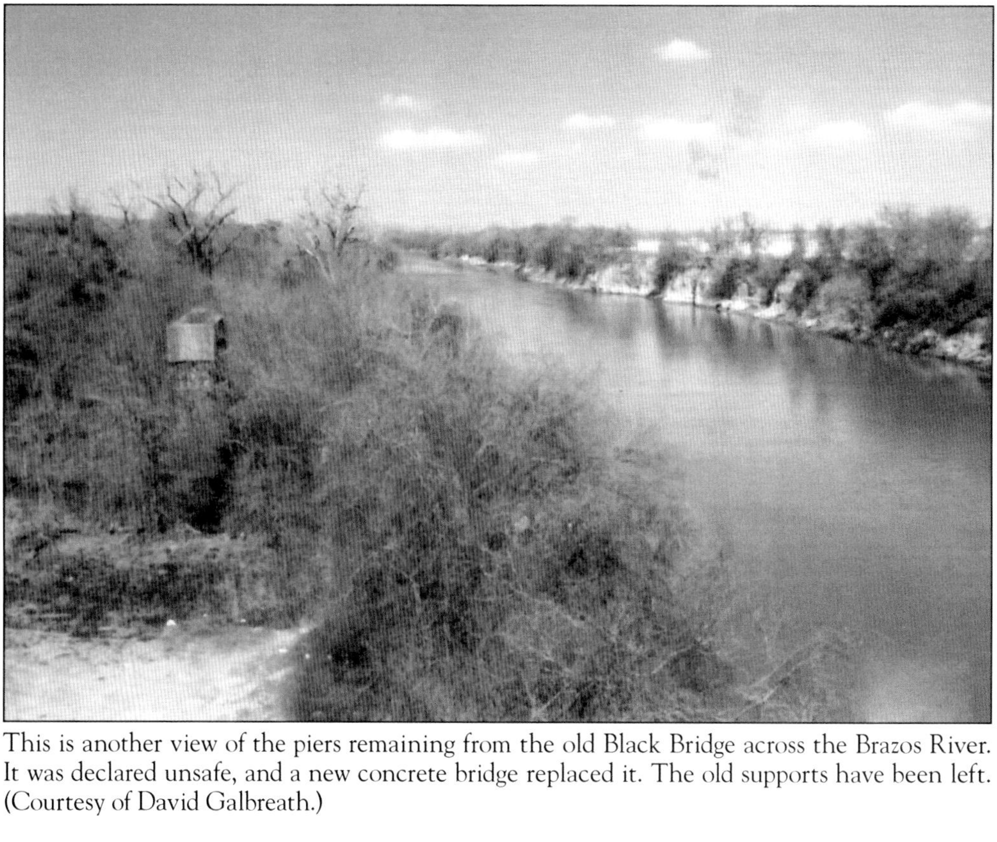

This is another view of the piers remaining from the old Black Bridge across the Brazos River. It was declared unsafe, and a new concrete bridge replaced it. The old supports have been left. (Courtesy of David Galbreath.)

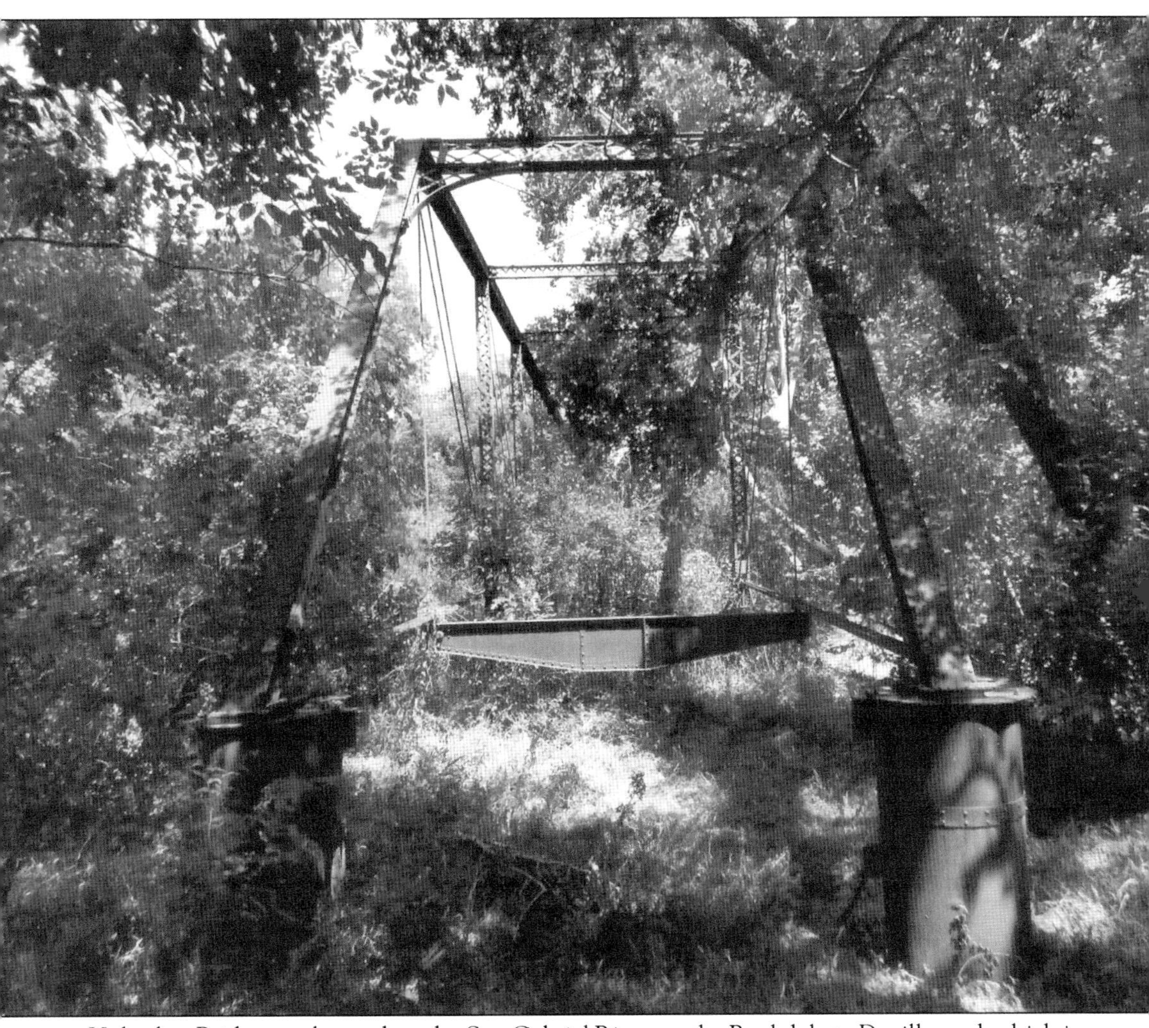

Holtzclaw Bridge was located on the San Gabriel River on the Rockdale to Davilla road, which is Farm to Market 487. The bridge was in bad shape in 1881, so Holtzclaw sold it for $500. (Courtesy of Jack and Beth Brooks.)

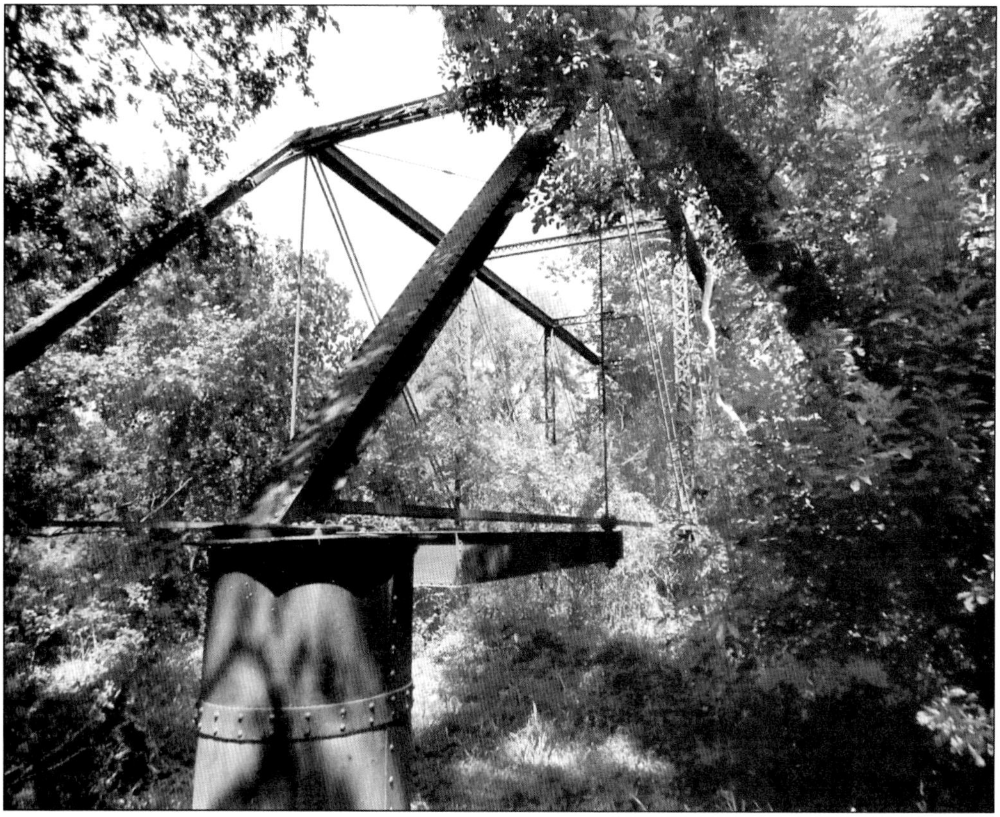

The International & Great Northern Railroad (IG&N) made repairs on the Holtzclaw Bridge and made it usable again. Milam County paid the IG&N $500 for its services. (Courtesy of Jack and Beth Brooks.)

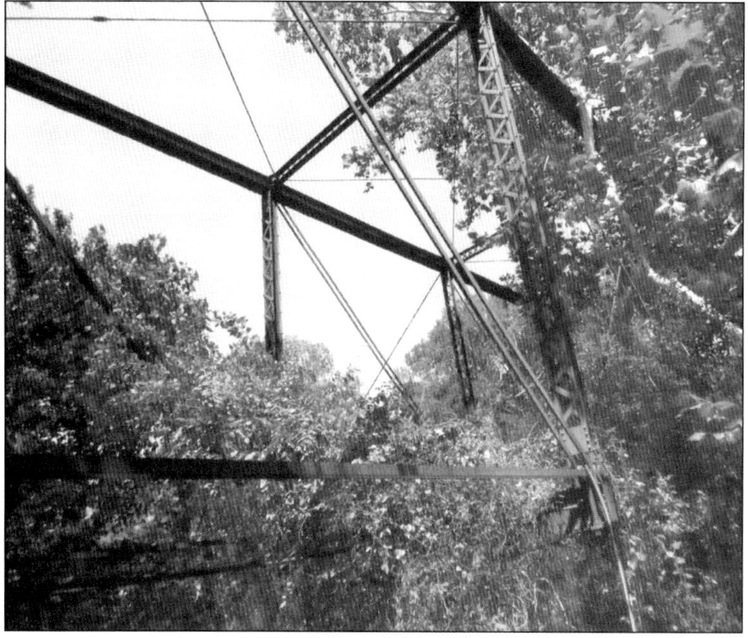

A new Holtzclaw Bridge was constructed in 1886 by the Kansas City Bridge and Iron Company. The company received $3,500 for its work. (Courtesy of the Milam County Historical Museum.)

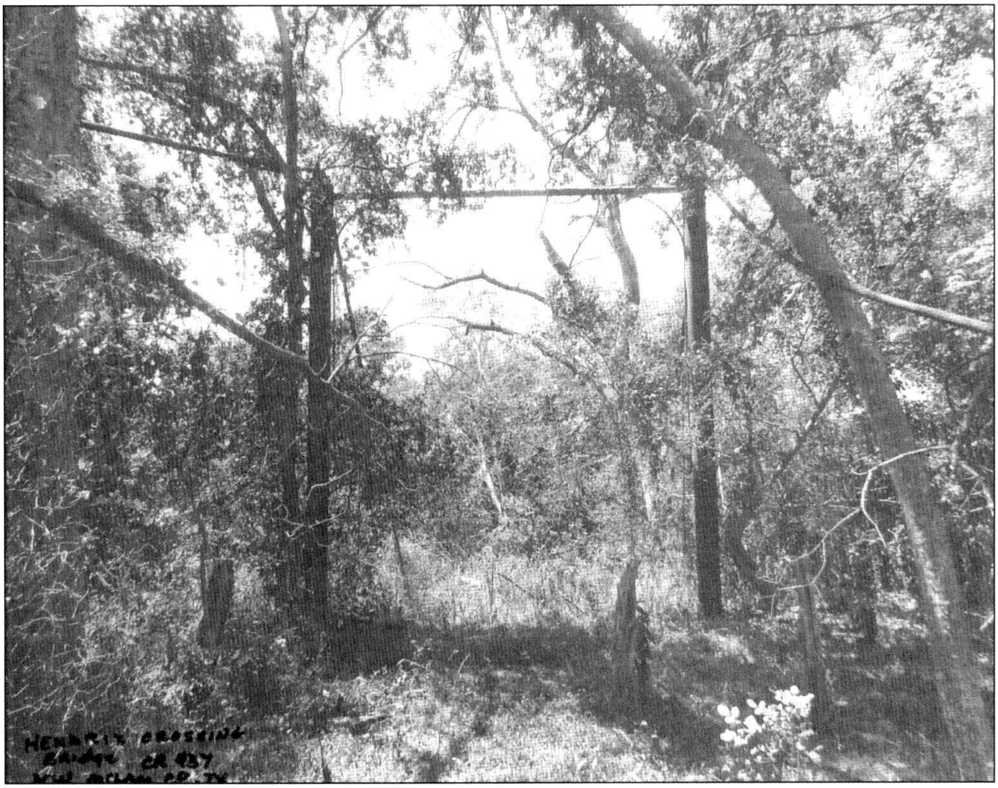

The above photograph shows Hendrix Crossing Bridge, which was located south of the city of Rogers in the westernmost part of Milam County. It was a cable bridge across the Little River. At right is another photograph of Hendrix Crossing Bridge. (Above, courtesy of David Galbreath.)

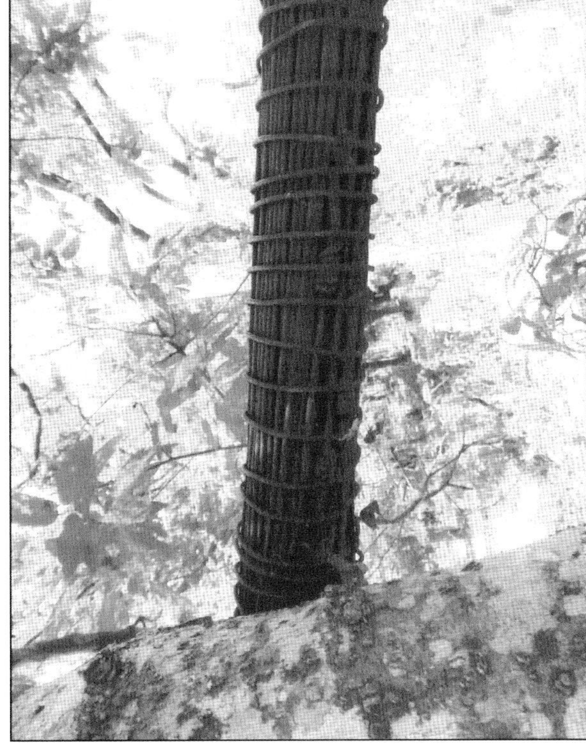

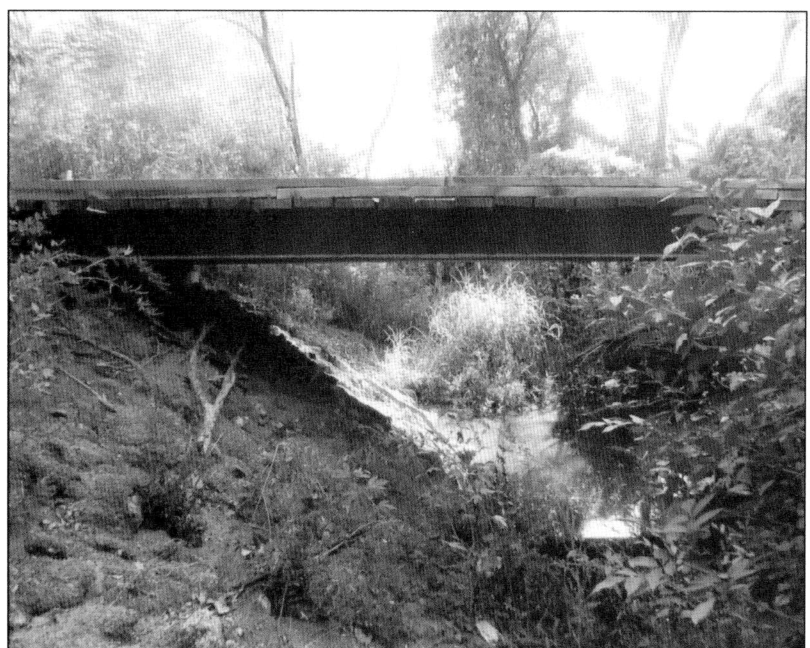

Table Bridge is one of the lesser-known bridges in Milam County. It is located 3.4 miles south of Farm to Market 845 near Cameron. (Courtesy of Jack Brooks.)

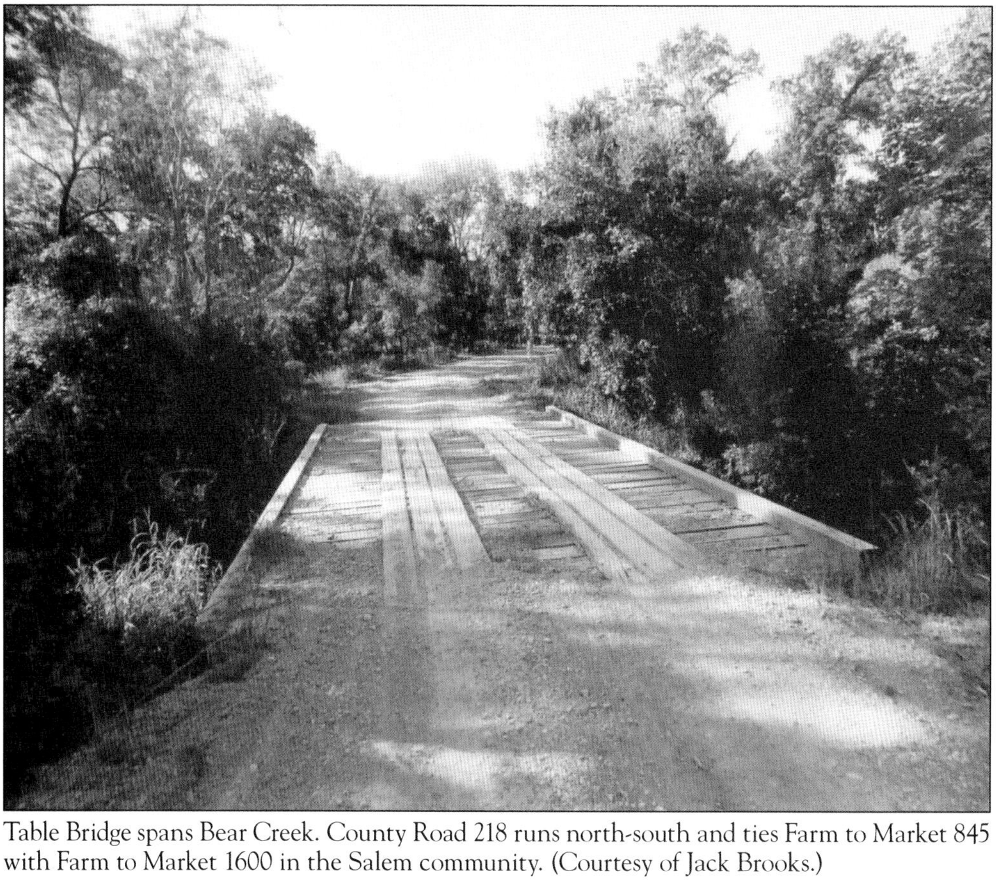

Table Bridge spans Bear Creek. County Road 218 runs north-south and ties Farm to Market 845 with Farm to Market 1600 in the Salem community. (Courtesy of Jack Brooks.)

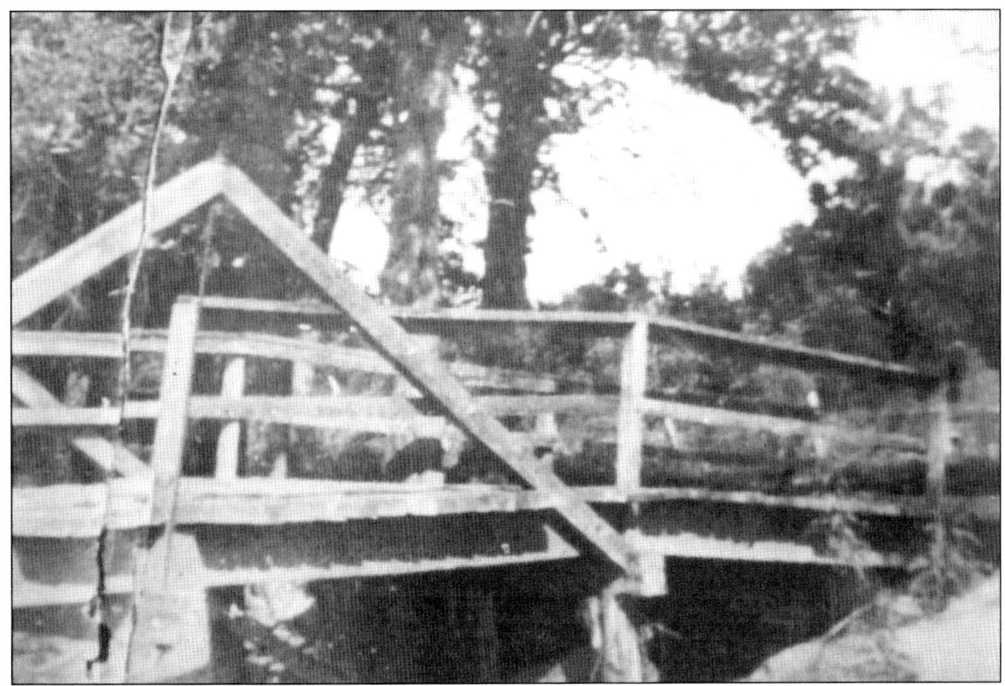

Cedar Creek Bridge was built over the creek of the same name, which rises four miles south of Milano in southeastern Milam County. During the early 1840s, just prior to the Republic of Texas days, there were many skirmishes around this area as the settlers tried to drive the Indians away from the land. (Courtesy of the Milam County Historical Museum.)

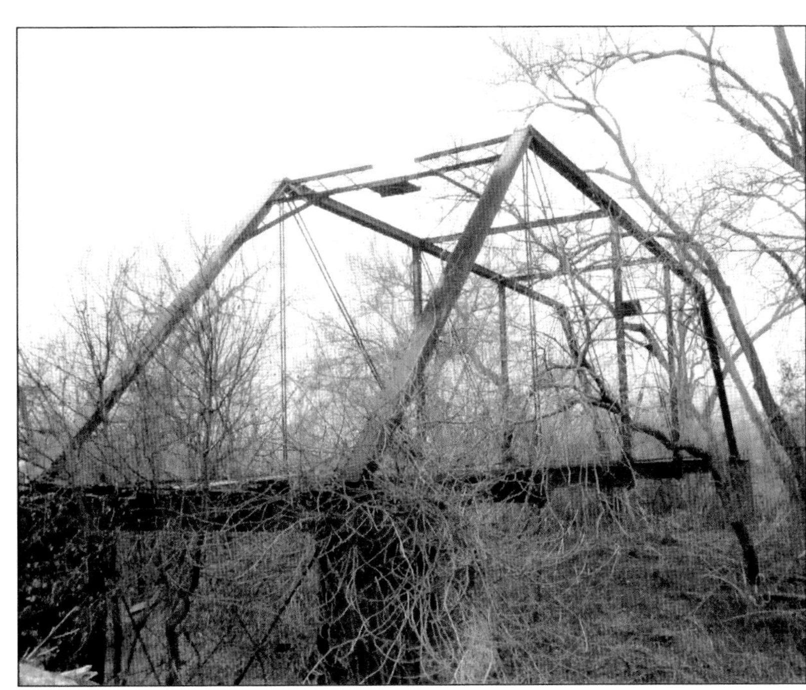

Unknown Bridge crosses Brushy Creek about two miles south of the intersection of Highway 79 and Farm to Market 486 in Thorndale. (Courtesy of Jack Brooks.)

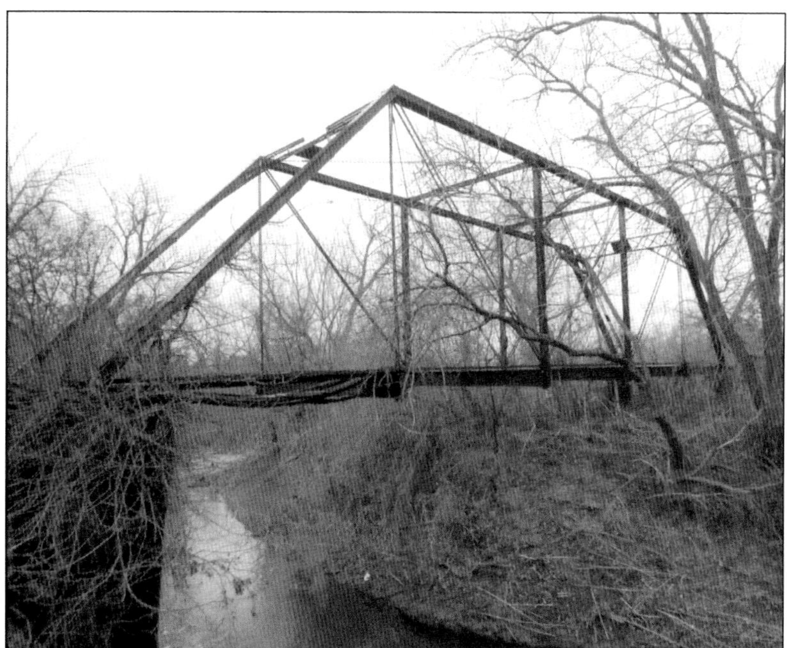

Unknown Bridge is now located in a stand of trees and hidden by heavy growth. There are no longer any approach ramps. (Courtesy of Jack Brooks.)

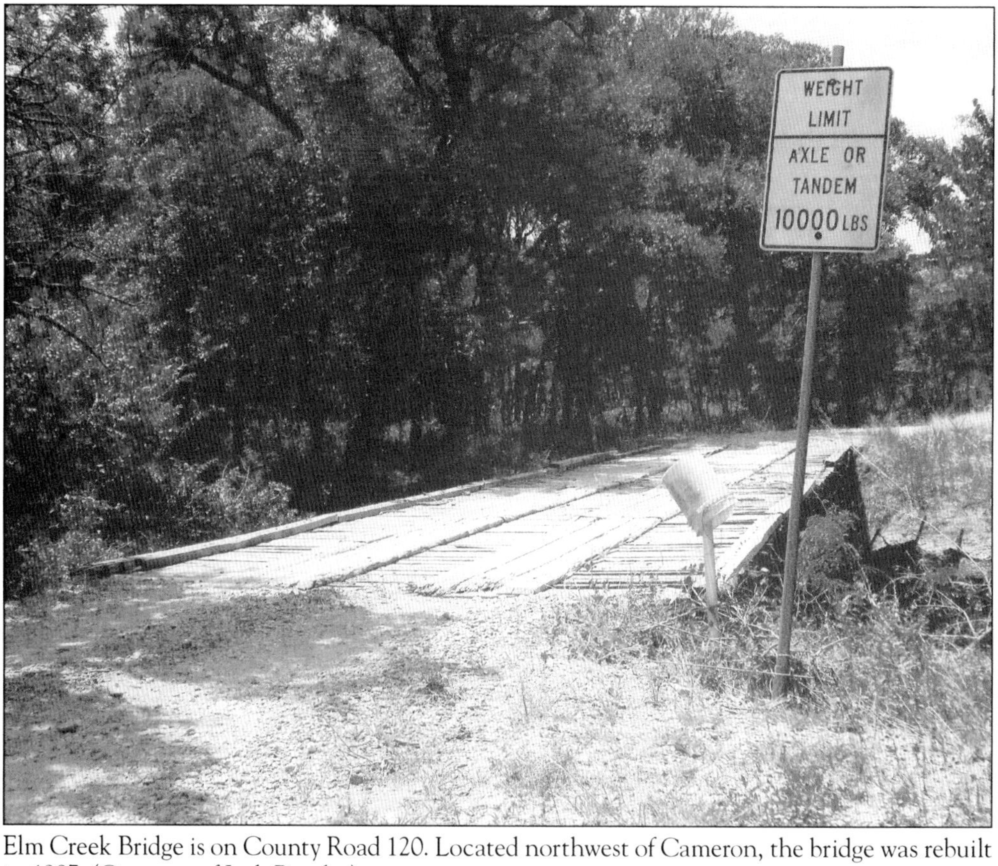

Elm Creek Bridge is on County Road 120. Located northwest of Cameron, the bridge was rebuilt in 1987. (Courtesy of Jack Brooks.)

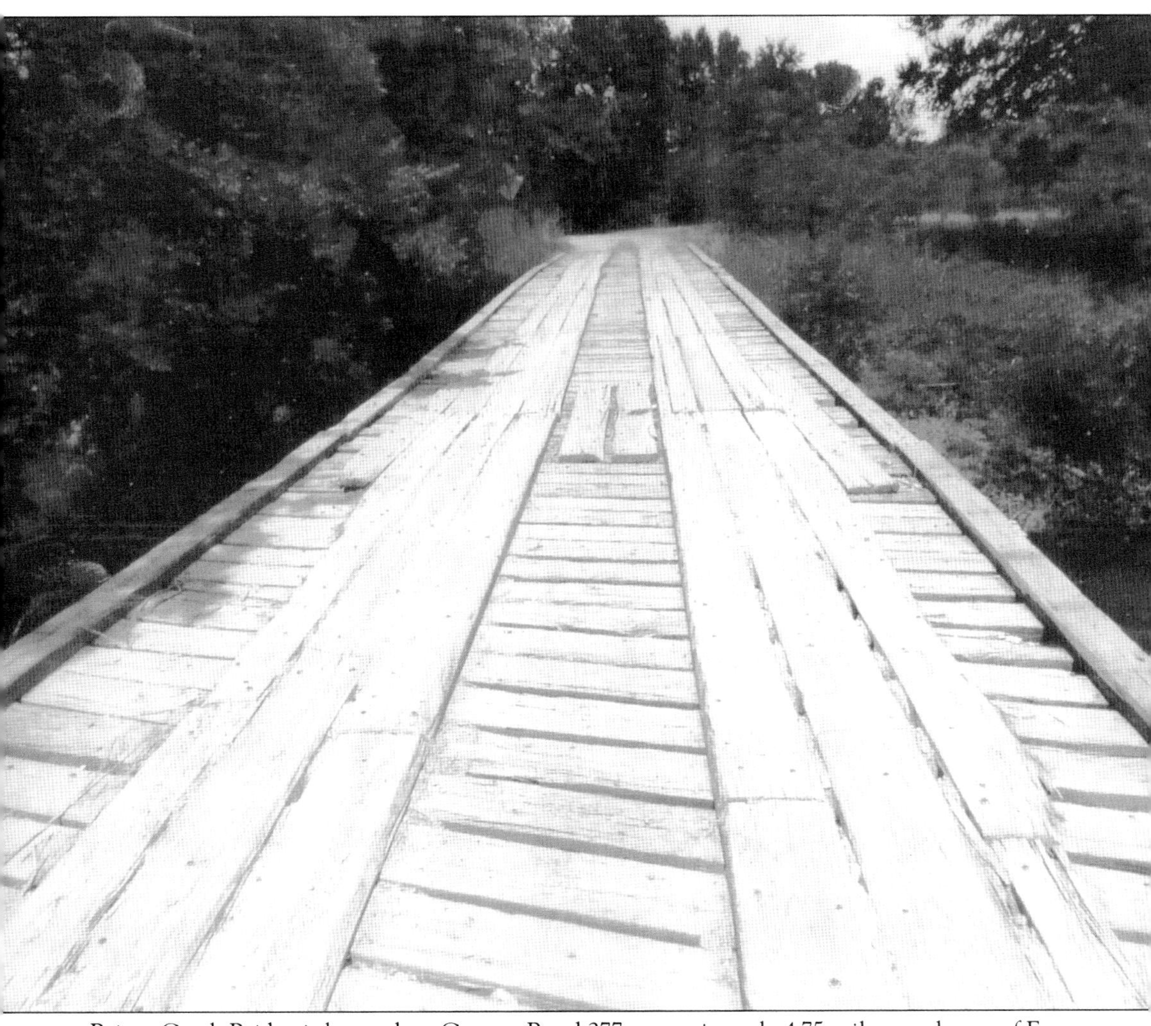

Briary Creek Bridge is located on County Road 277 approximately 4.75 miles northwest of Farm to Market 2027. Its primary purpose is to carry highway traffic over the waterway. Named for the creek, Briary itself is a ghost town near Rosebud. (Courtesy of David Galbreath.)

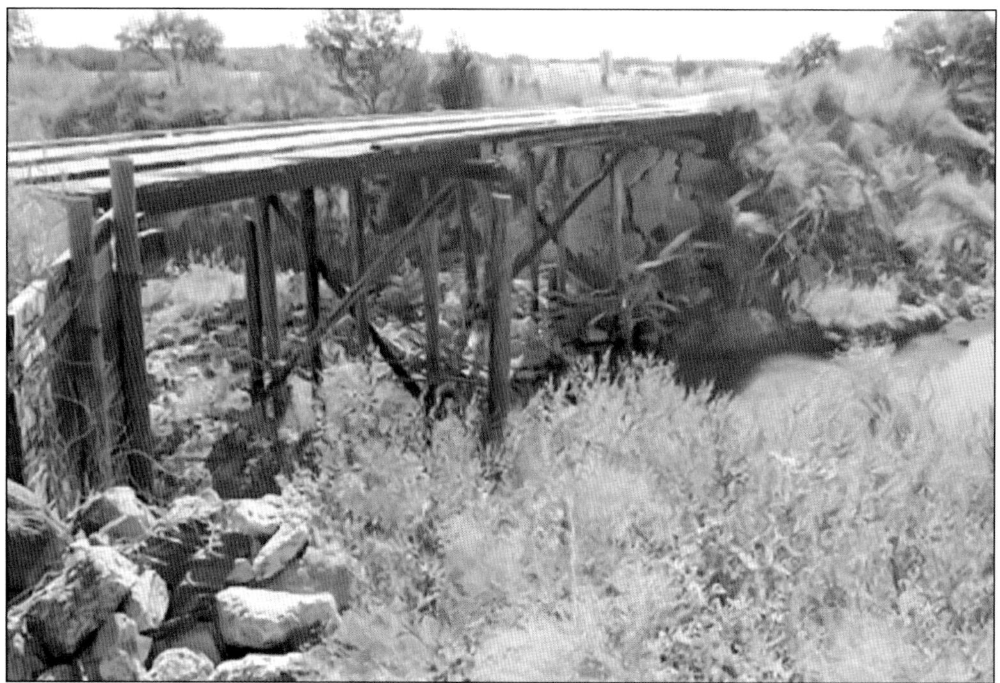

2491115 In this photograph is Little Pond Creek Bridge as it appeared on April 17, 2016. The bridge had been damaged by rains previously, but it was considered to be in satisfactory condition and safe for vehicles to cross. (Courtesy of Donald Shuffield.)

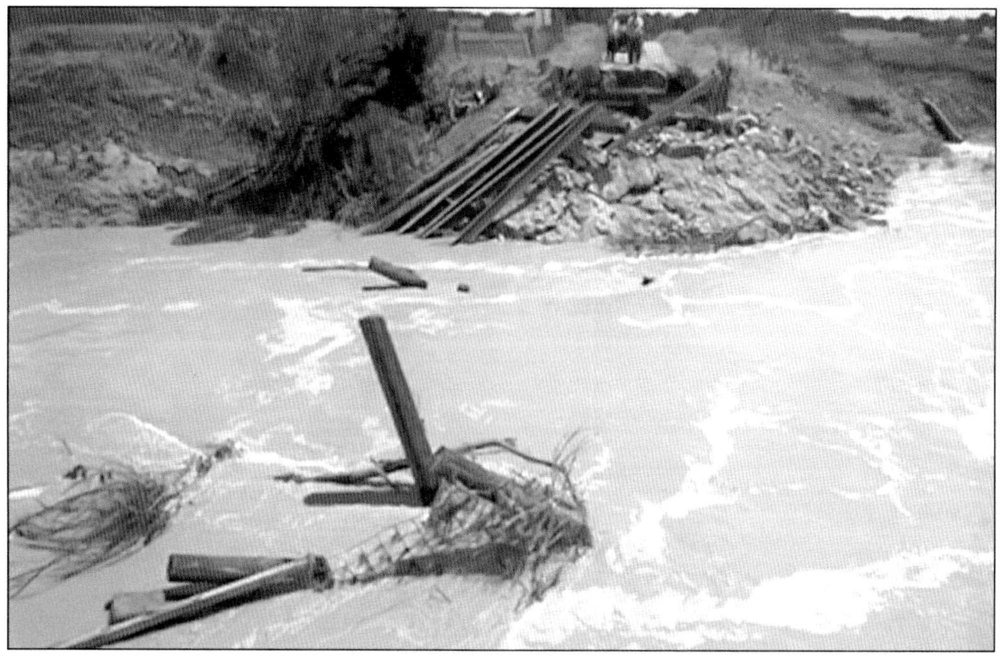

Persons planning to travel County Road 267 on April 18, 2016, were surprised to find that the 83-foot-long bridge was gone. This bridge was washed away due to the unseasonable amount of rain dumped into Little Pond Creek. The creek's flow increased from 0 to 20,000 cubic feet per second, resulting in the catastrophic effects seen in this photograph. (Courtesy of Donald Shuffield.)

Five

"Ramblin' Round"

The bridges of this area and their ongoing histories provide a colorful and fascinating insight into the history of Milam County itself. As Milam County historian and author Marion Travis once wrote, "First there was the land, then came the people." And come they did—from all over. As they came, they embraced the land. As they interacted with others, they became intertwined in the story of this county. They were authors, artists, historians, and just plain folks who came to live, work, and raise families. As families rambled around Milam County, enjoying the scenery and the people, they viewed the bridges on Sunday afternoons, picnicked on the spans, and posed for family photographs of memorable occasions there, and as they did these things, bridges became a part of their memories. This is continuing today as schoolchildren visit Apache Pass and other sites along the El Camino Real de lost Tejas and experience the wonder of history as a living thing.

Included in this section is the image of a GPS map to help visitors get from one site to the next. Traces of some of the bridges are very hard to find, but with the map, people should be able to locate them.

Note that the phrase "Ramblin' Round' was used for many years by W.H. Cooke Sr., longtime editor of the *Rockdale Reporter.*

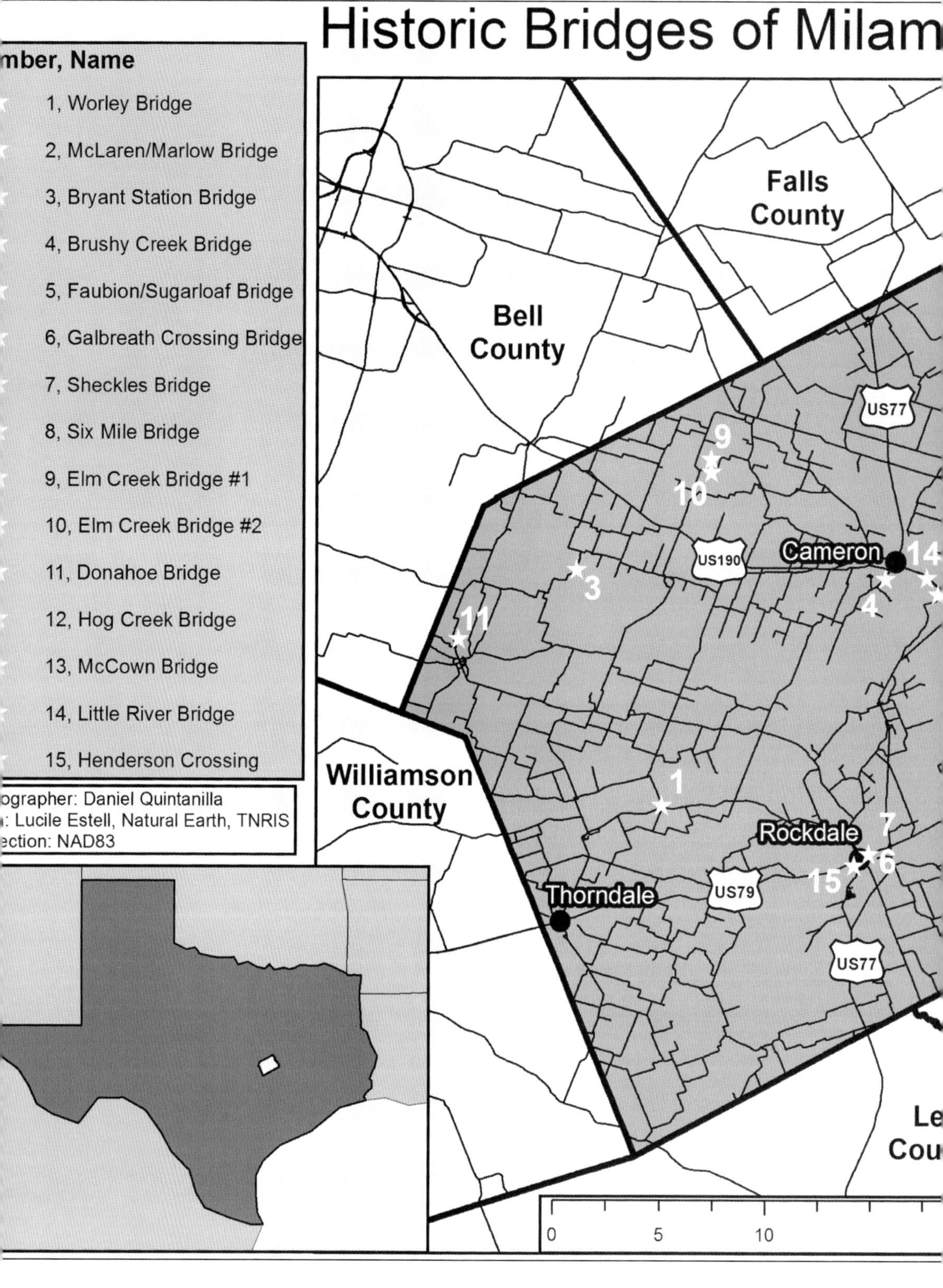
Historic Bridges of Milam
mber, Name
1, Worley Bridge
2, McLaren/Marlow Bridge
3, Bryant Station Bridge
4, Brushy Creek Bridge
5, Faubion/Sugarloaf Bridge
6, Galbreath Crossing Bridge
7, Sheckles Bridge
8, Six Mile Bridge
9, Elm Creek Bridge #1
10, Elm Creek Bridge #2
11, Donahoe Bridge
12, Hog Creek Bridge
13, McCown Bridge
14, Little River Bridge
15, Henderson Crossing
ographer: Daniel Quintanilla
: Lucile Estell, Natural Earth, TNRIS
ection: NAD83
Falls County
Bell County
Williamson County
US77
US190
Cameron
Rockdale
Thorndale
US79
US77
0
5
10

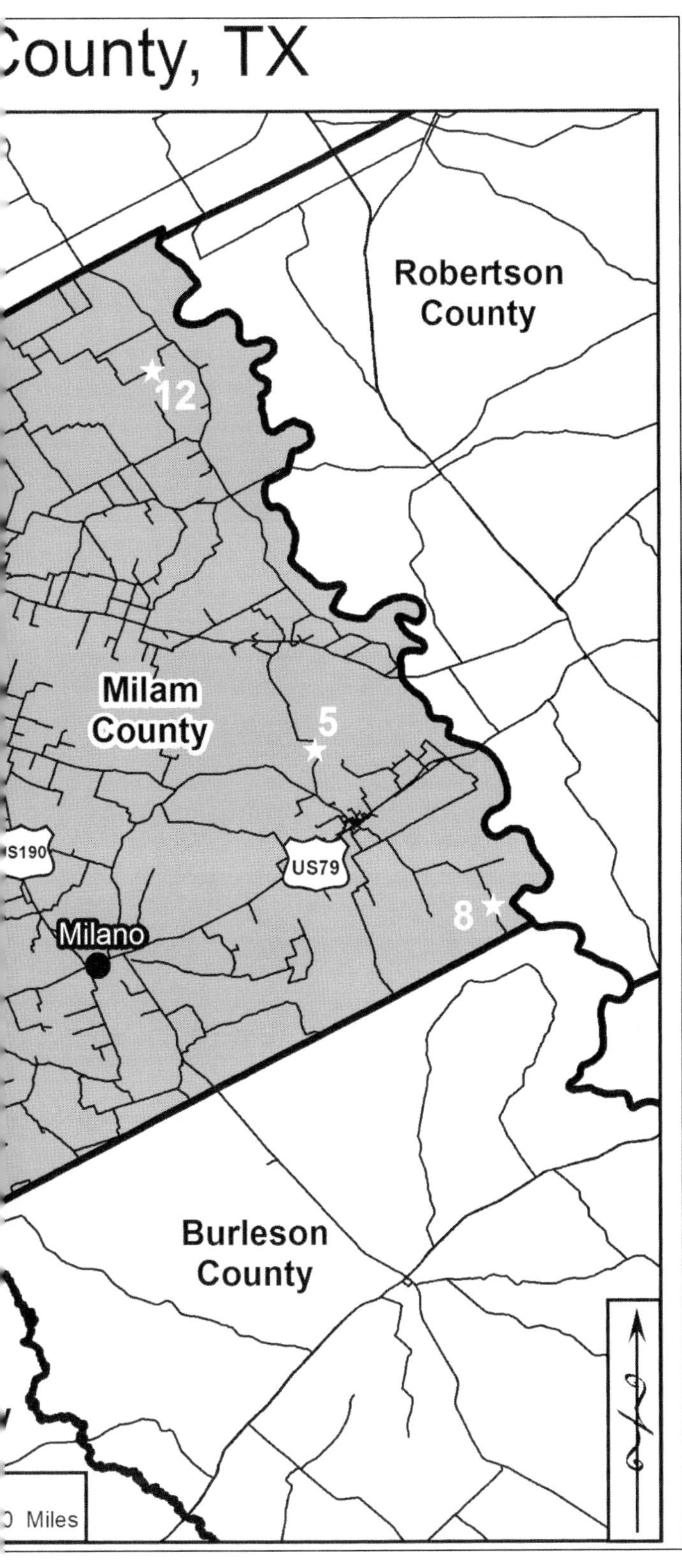

This map showing the locations of many of the bridges is included for those interested in visiting these sites. (Courtesy of Daniel Quintanilla.)

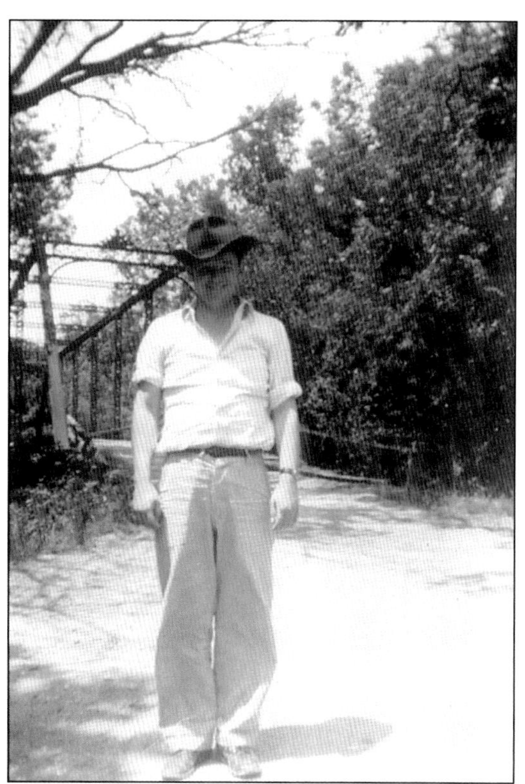

Milam County resident Harold Pruett stands in front of the historic Worley Bridge in 1947. (Courtesy of John Pruett.)

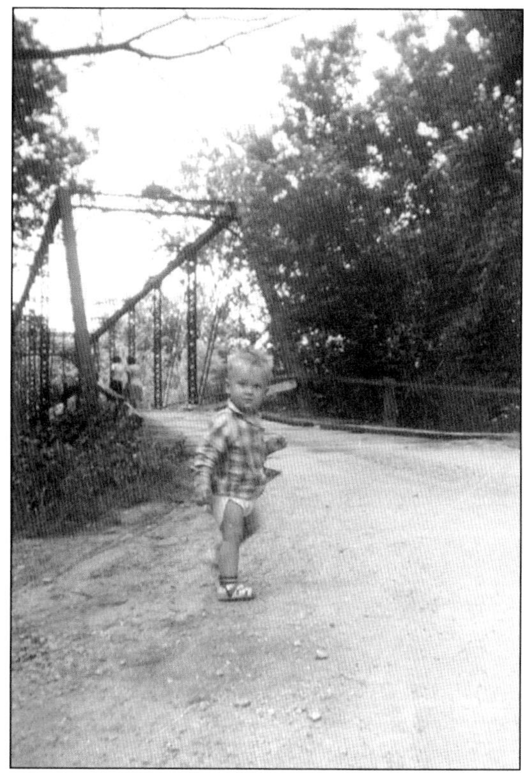

Harold's son, John Pruett, is pictured at the same bridge in 1947. (Courtesy of John Pruett.)

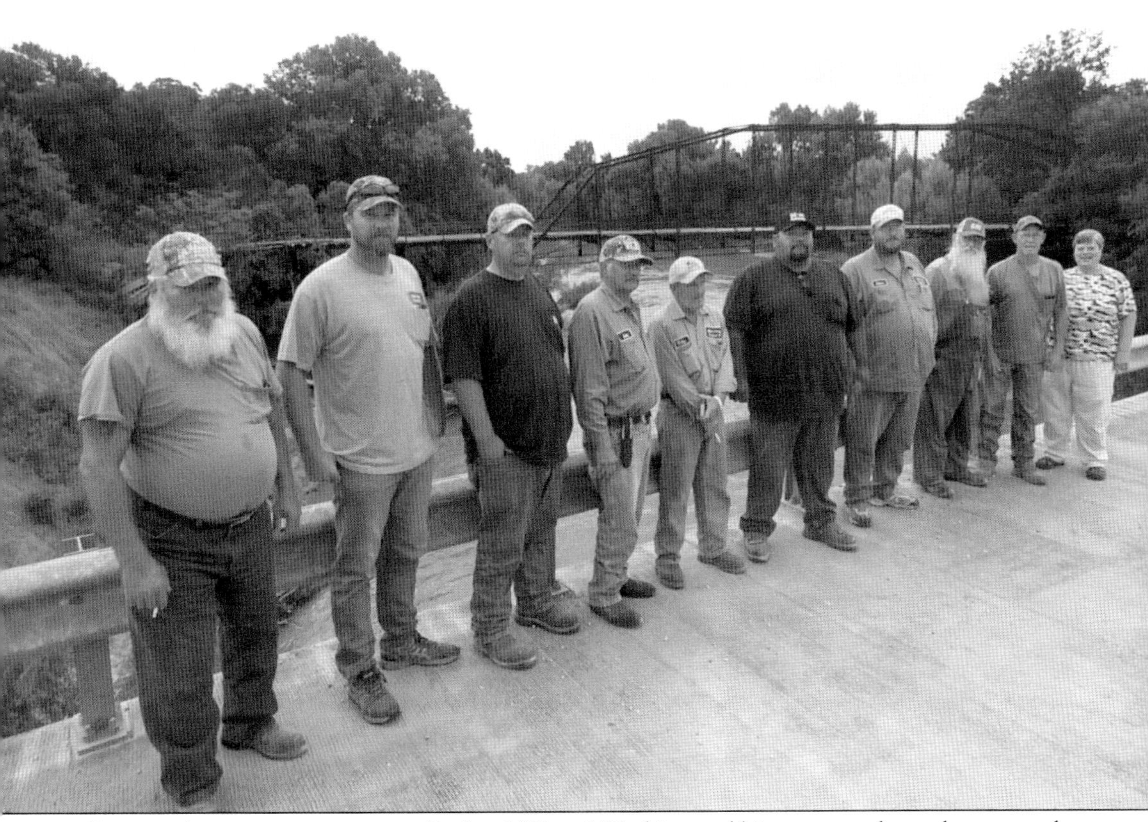

Precinct 1 county commissioner Richard "Opey" Watkins and his crew are shown here near the remains of the Bryant Station Bridge. From left to right are Elbert Svetlik, Trot Tucker, George Junek, Leon Hollas, Billy Allen, David Trevino, Calvin Junek, James Junek, Commissioner Watkins, and Aline Junek. (Courtesy of Jack Brooks.)

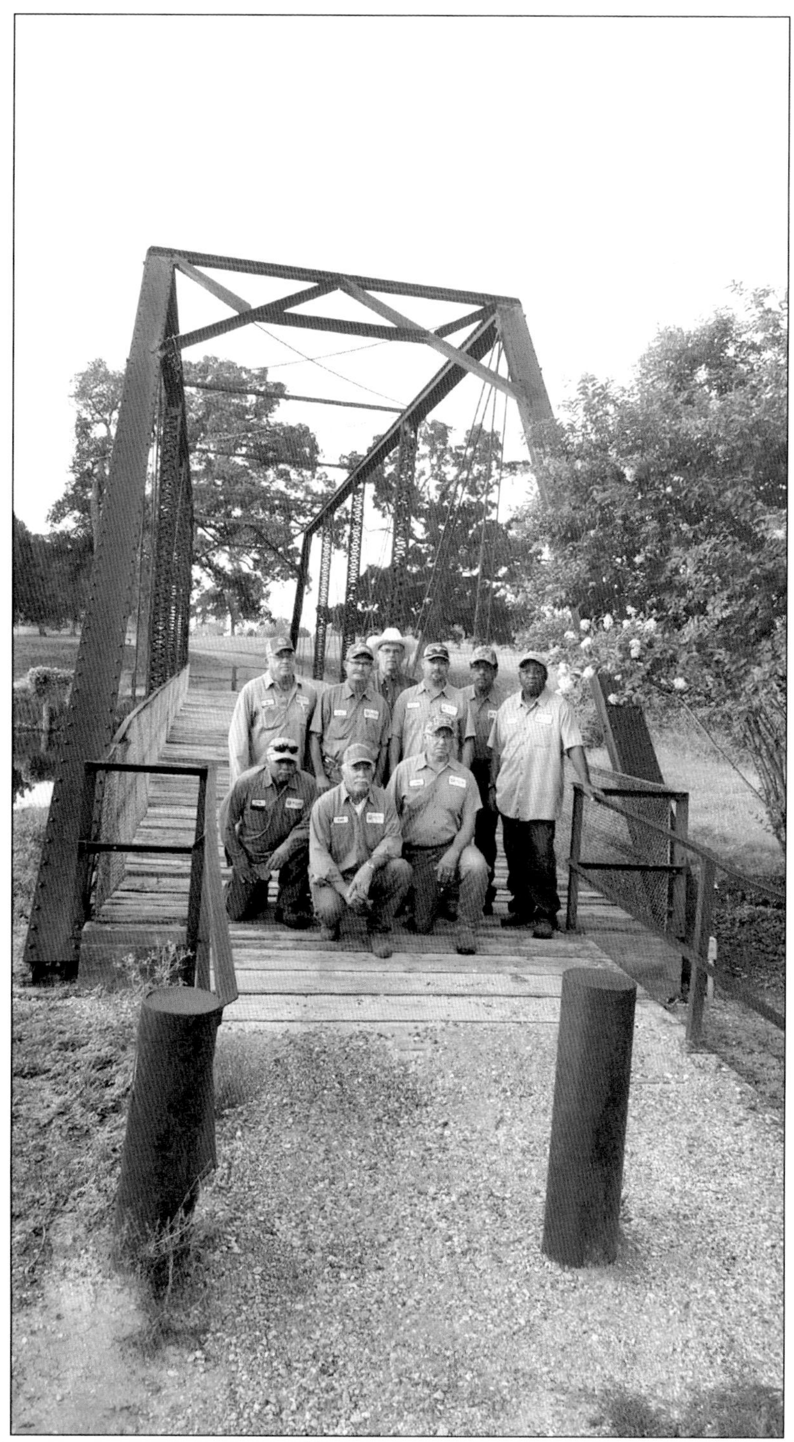

Precinct 2 county commissioner Donald Shuffield and his crew are pictured here. From left to right are (first row) Cruz Reyes, Scott Mitchan, and Leroy Vonsa; (second row) Ed Dohnalik, Joe Vaculin, Commissioner Donald Shuffield, Clayton Gann, Cliff Roseman, and Danny Smith. (Courtesy of Geri Burnett.)

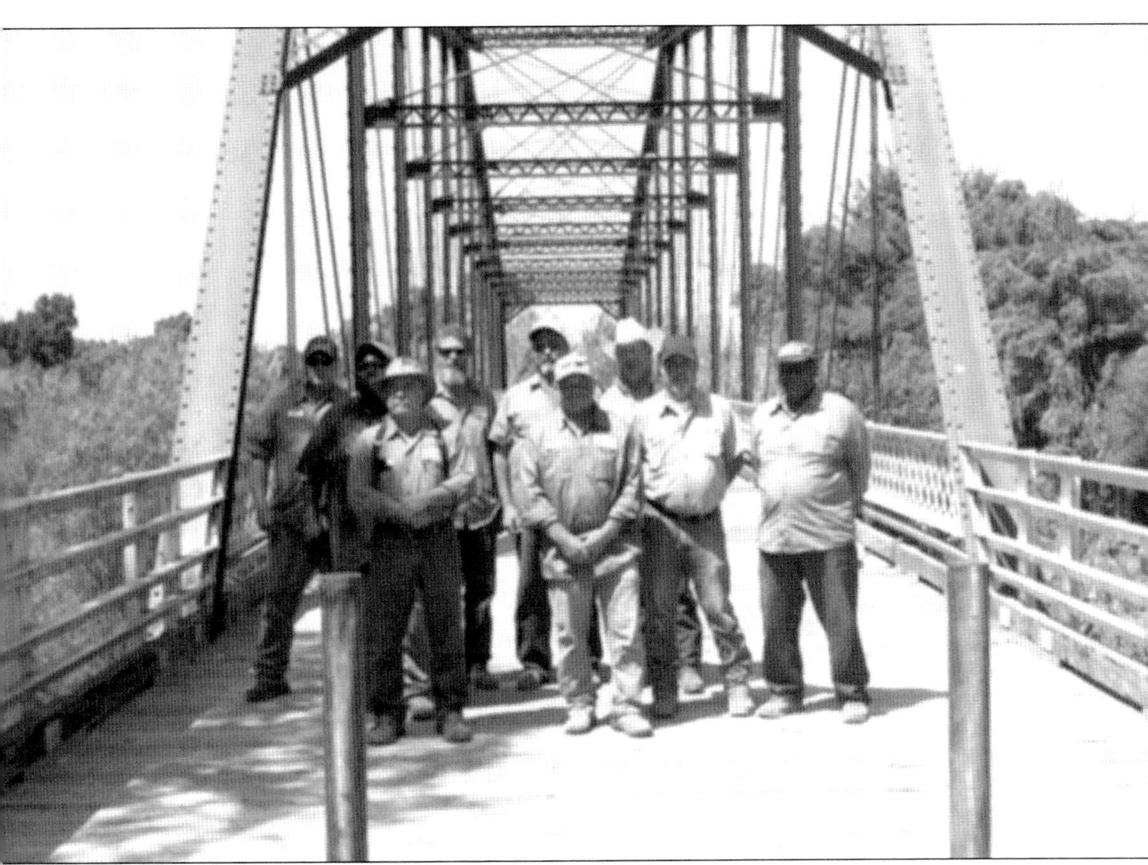

This photograph shows Precinct 3 county commissioner John Fisher and his crew at the Faubion Bridge. From left to right are Jeff Gebhart, Nigel Crawford, Bob Morrison (foreground), Steven Honeycutt, Kerry Schlemmer, David Head (foreground), Commissioner Fisher, Mike Harris, and Bennie Swift. (Courtesy of John and Brenda Pruett.)

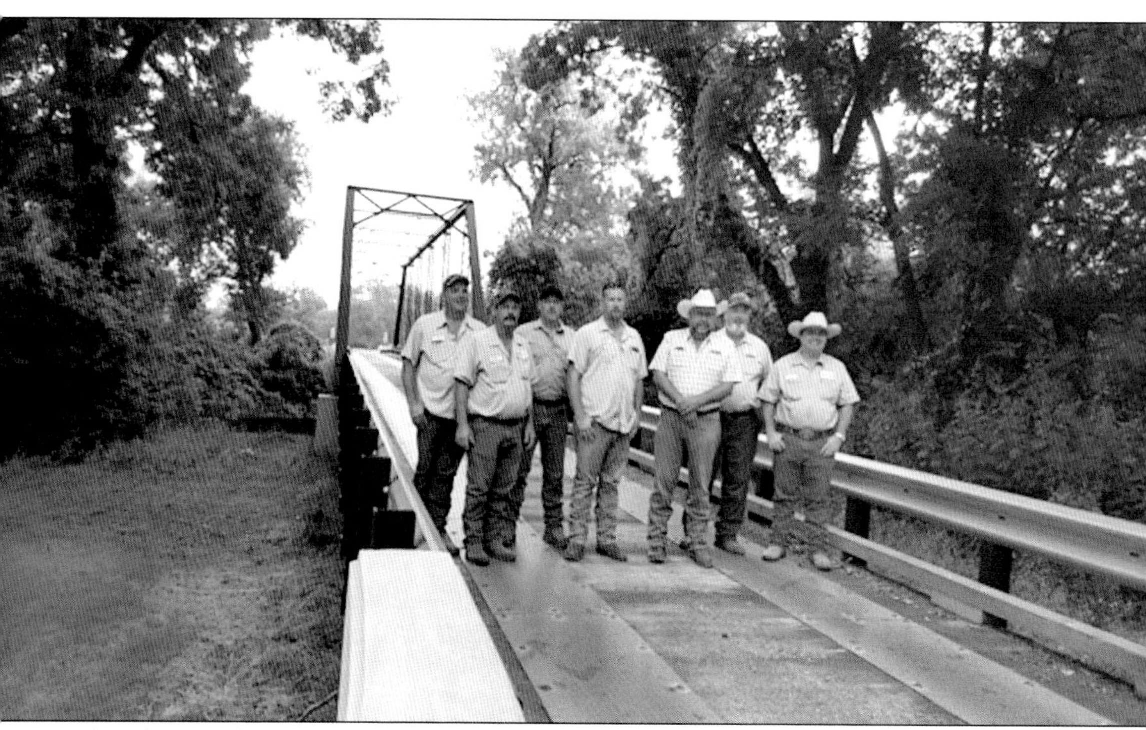

This photograph is of Precinct 4 county commissioner Jeff Muegge and his crew. From left to right are Johnny Hubnik, Billy Felfe, Travis Hoskins, Jeromy Hoskins, Commissioner Muegge, Russell Niemtschk, and Neil Caffey. (Courtesy of Geri Burnett.)

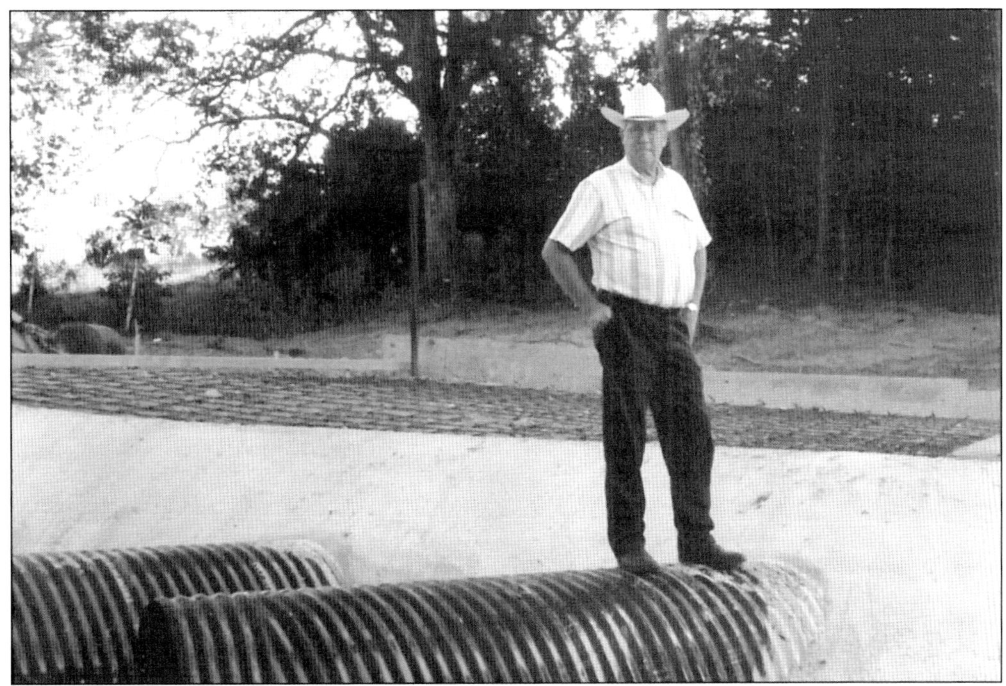

Former commissioner Troy Mode actively promoted the building of small bridges over low-water crossings. This was a great help to people in his precinct as they moved about the county. (Courtesy of Dolores Mode.)

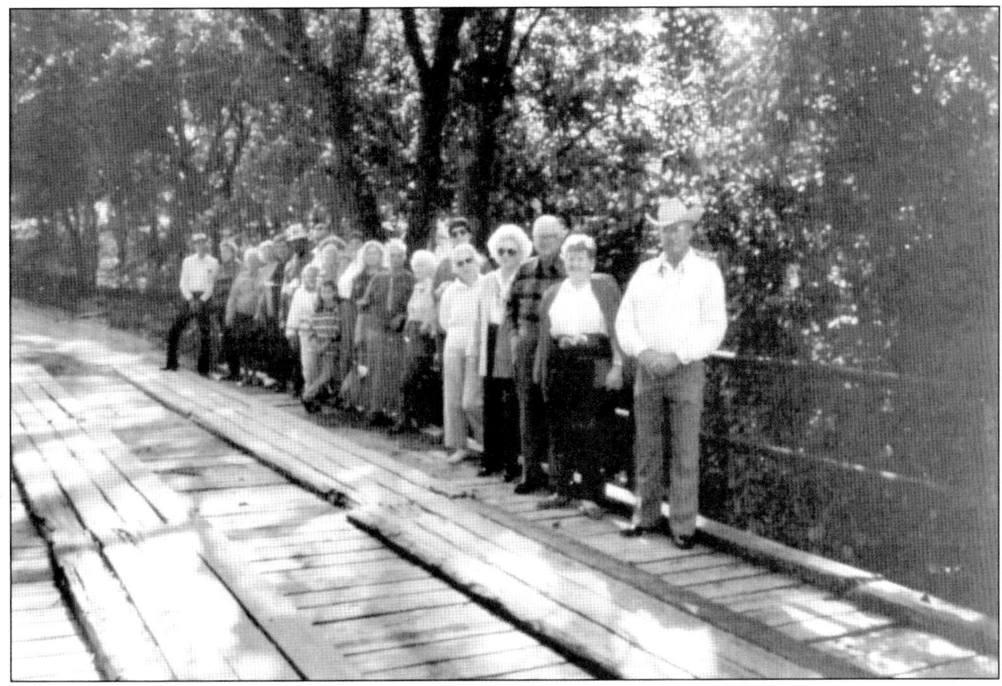

Milam County citizens have long been interested in the history of the county, especially in bridges. Identified in this photograph are Troy and Dolores Mode. (Courtesy of Dolores Mode.)

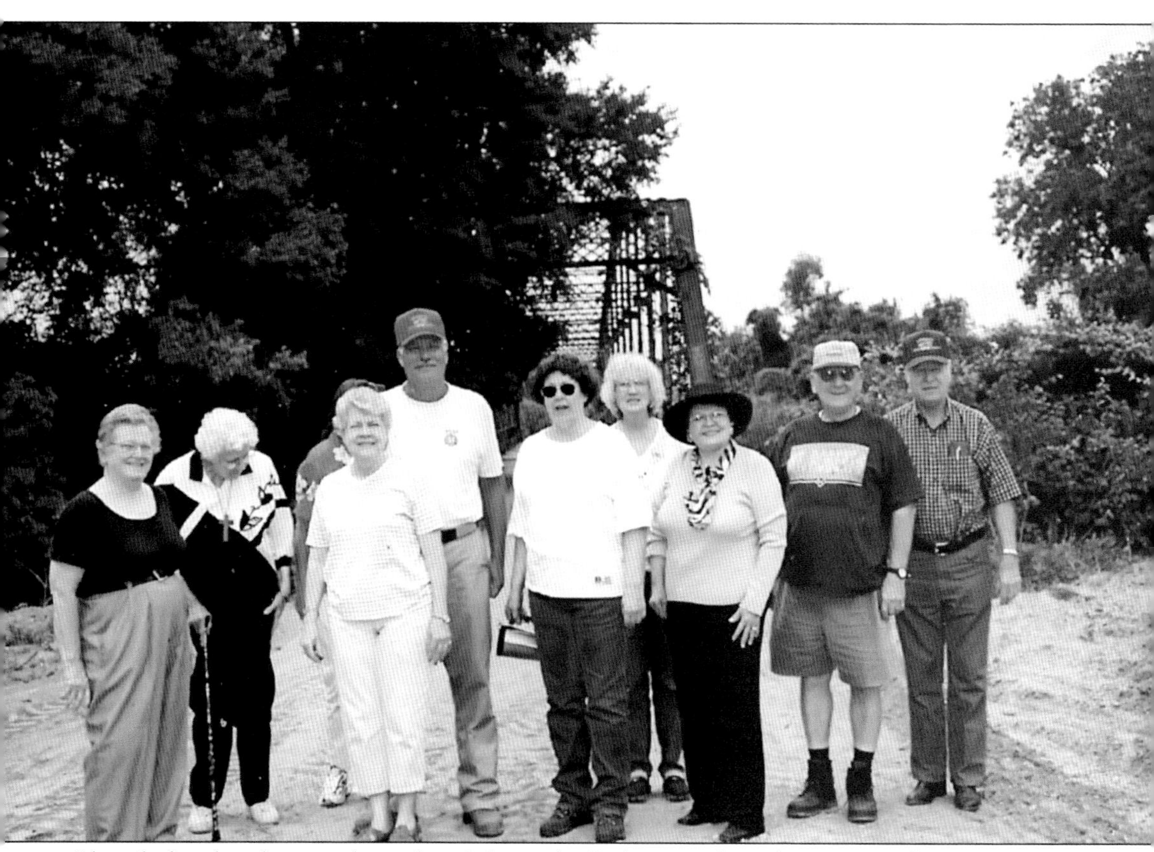

Identified in this photograph are, from left to right, Dolores Mode, Helen Perry, Willyne Stanislaw, Bert Dockall, Pat Watson, Joy Graham, and Marie Hubert. The two people at right are unidentified. (Courtesy of Dolores Mode.)

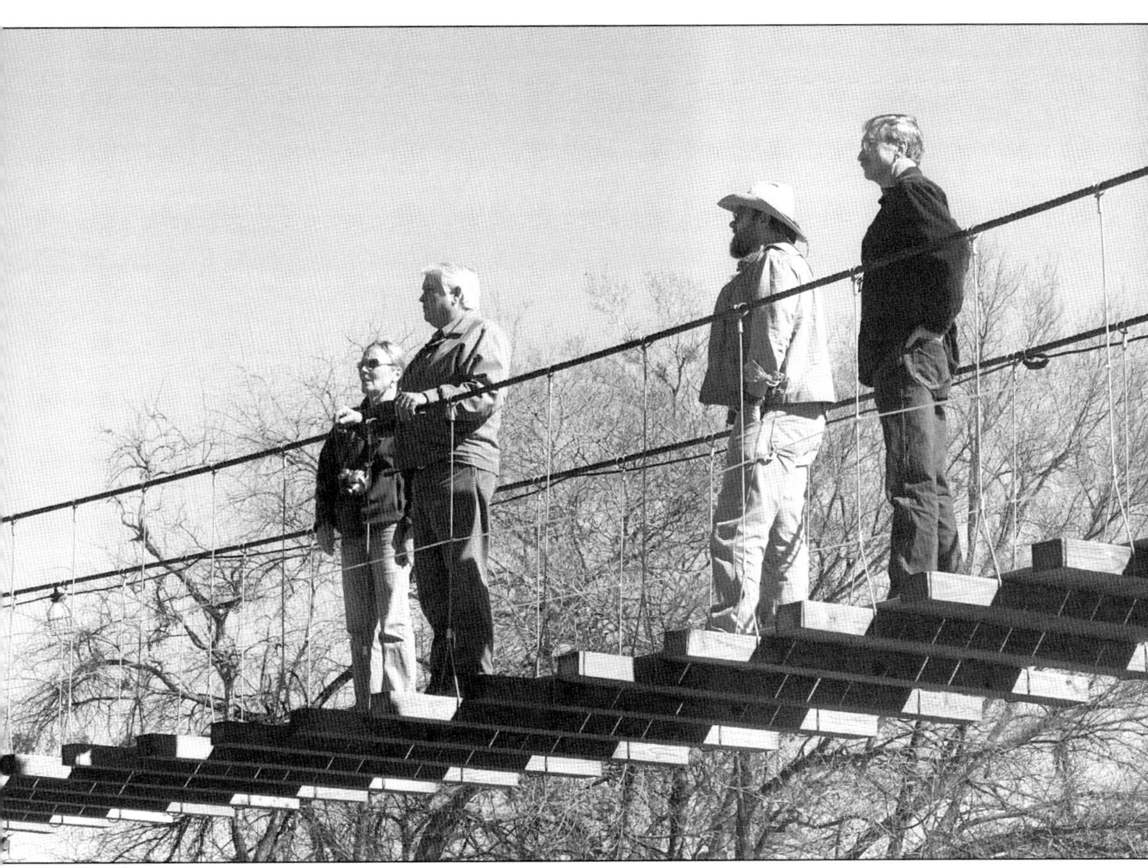

This is a photograph of the suspension bridge that spans a portion of Apache Pass. Identified on the bridge are, from left to right, Susan Boyle, retired historian from the National Park Service; Kit Worley, owner of Apache Pass; Jeff Williams of Nacogdoches; and Mike Taylor, from the National Park Service. (Courtesy of Lucile Estell.)

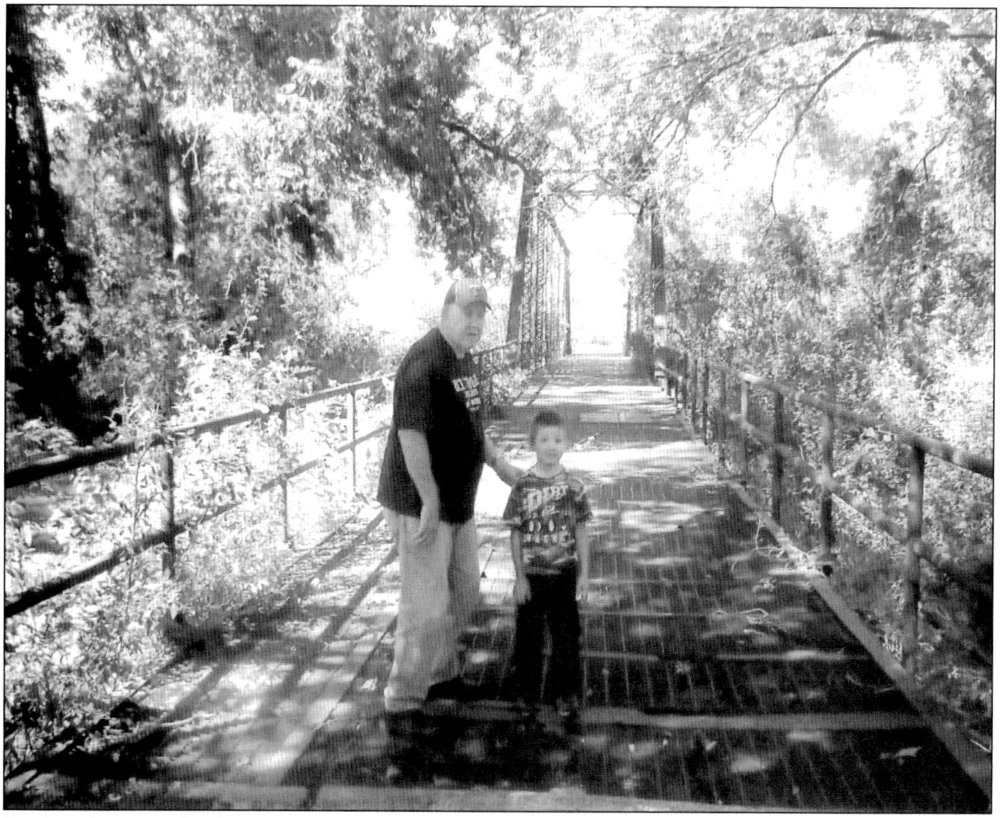

In the photograph above, David Galbreath and grandson Chance Allen Dempsey of Bullard, Texas, are seen at Worley Bridge on the San Gabriel River in 2011 before its restoration. Pictured below, on July 14, 2014, Dempsey revisits Worley Bridge with his grandfather to observe how it has changed. (Both, courtesy of Tracy Galbreath-Young.)

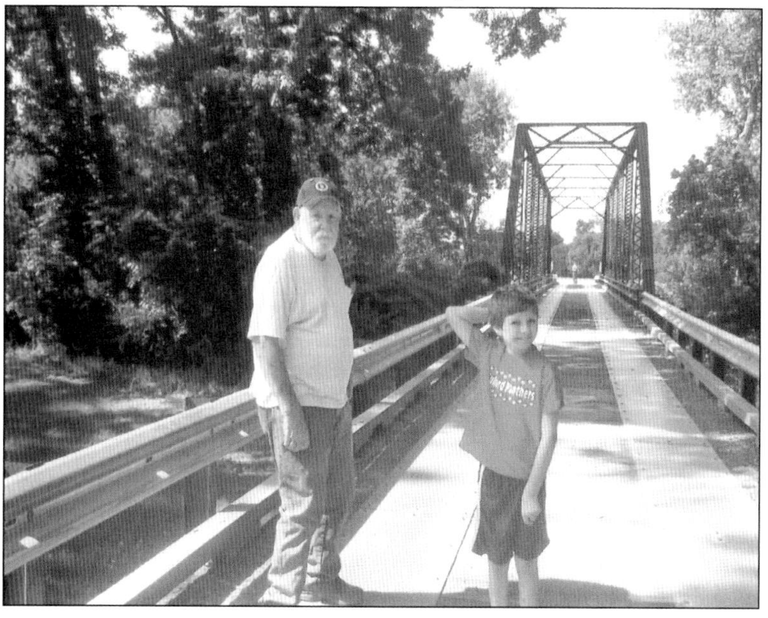

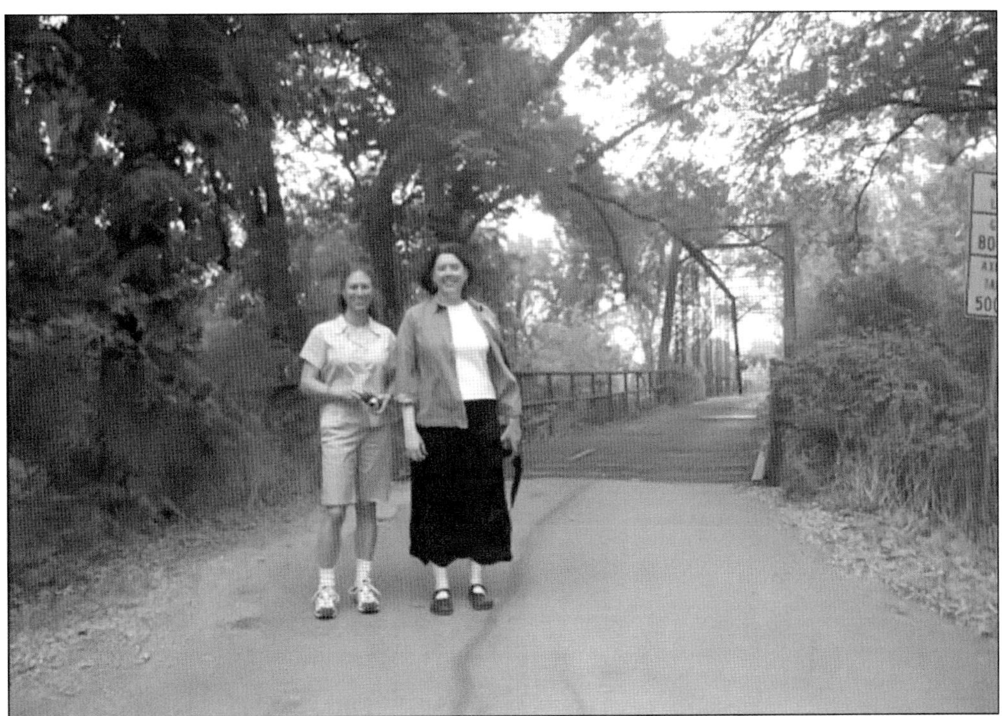

Bratten Thomason (left), of the Texas Historical Commission, and Rebecca Schwendler, of the National Trust for Historic Preservation, toured Worley Bridge as a part of their study of historic trail El Camino Real de los Tejas. (Courtesy of Geri Burnett.)

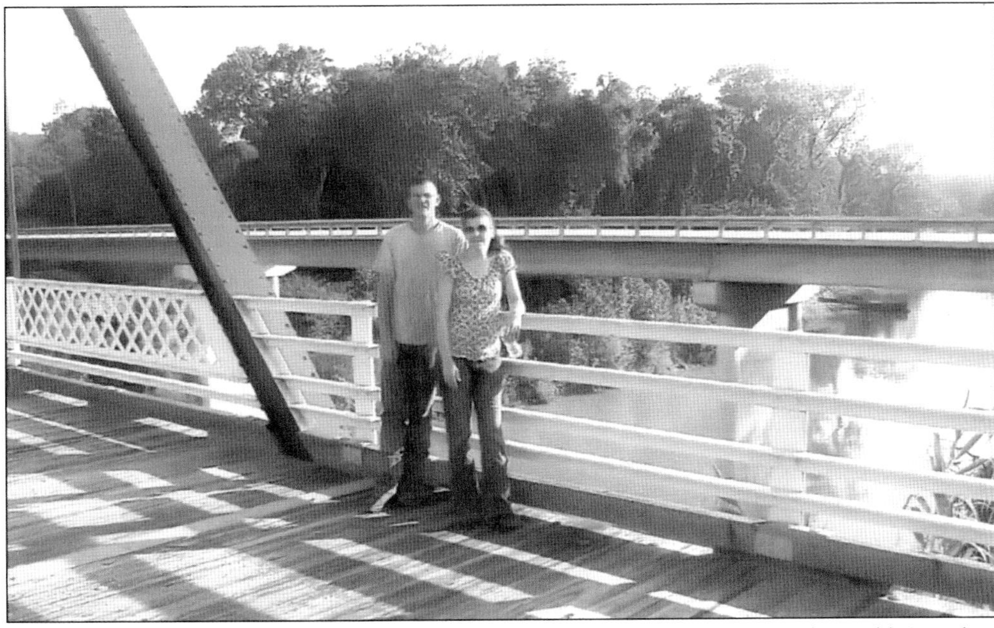

Before marrying Joshua Young, Tracy Galbreath took him to her favorite spot in the world, Sugarloaf Mountain and the Faubion Bridge in southeast Milam County. This bridge and mountain have been a traditional family-outing destination since her grandmother Tena Galbreath was a child. (Courtesy of David Galbreath.)

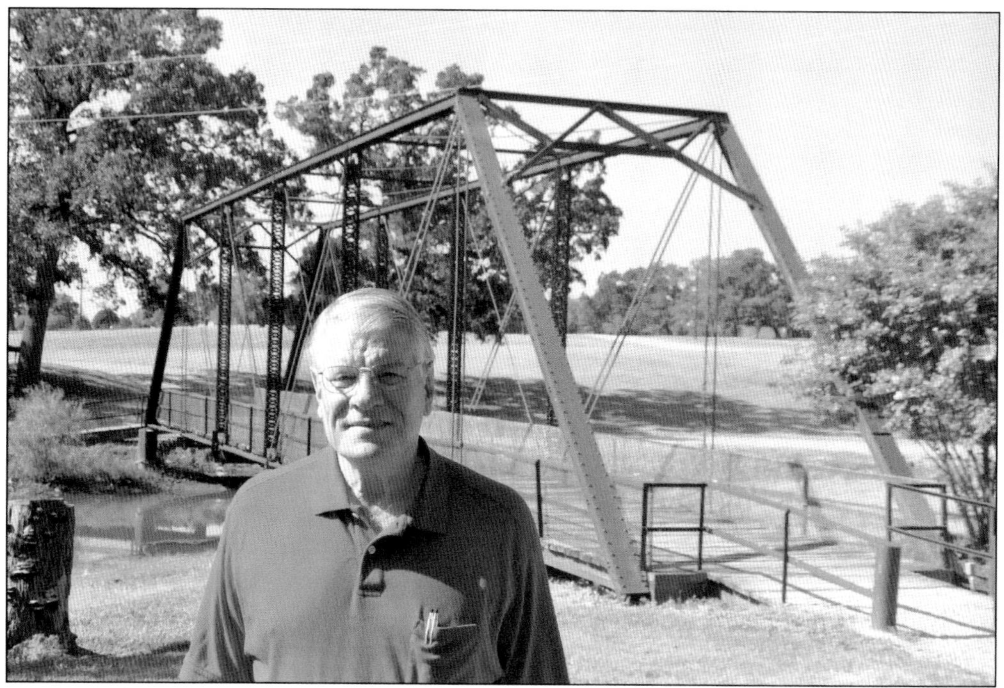

County Judge David Barkemeyer is the chairman of the Milam County Commissioners Court. He stands here in front of a historic bridge that has been moved from Brushy Creek to Ledbetter Park. (Courtesy of Chris Whittaker.)

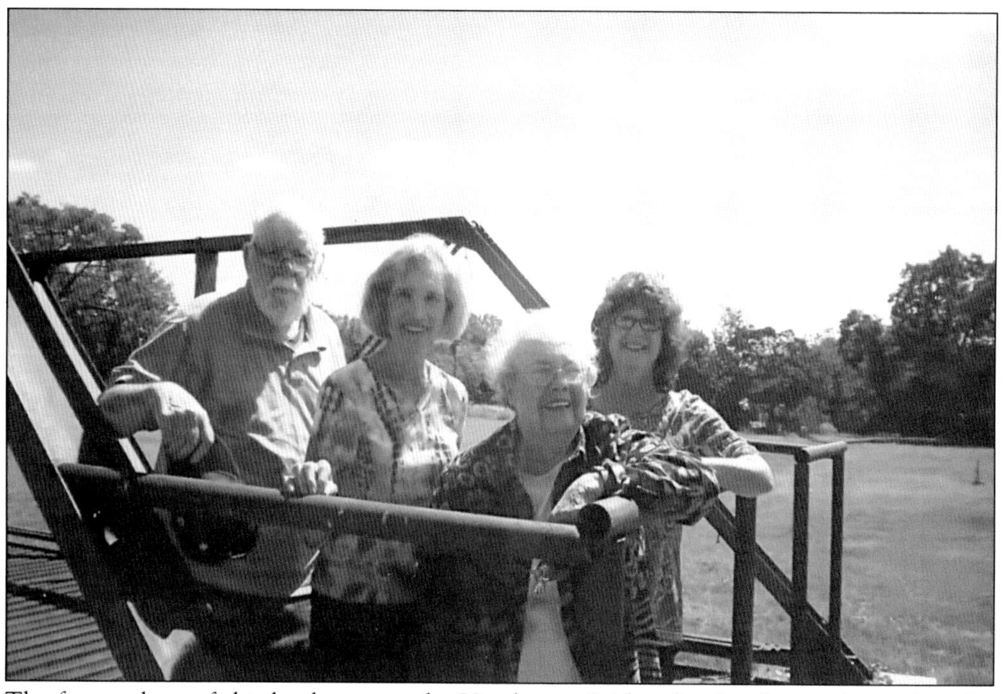

The four authors of this book pose at the Henderson Bridge that has been relocated to the Rockdale Skate Park. From left to right are David Galbreath, Joy Graham, Lucile Estell, and Carolyn Temple. (Courtesy of Darlen Graham.)

Six

Bridge Park

Preservation of these historical wonders presents an important task. What happens when a bridge is simply set aside and a new one is built? The preservation-minded city manager of Rockdale, Chris Whittaker, acted upon the idea of placing the bridges in a park. The city owns the property, and it is a perfect place for a park since it is a floodplain and nothing else can be done with it. He located a prominent place near a major highway in Rockdale and engineered not one but two of the lovely old structures into their new home and announced that there would be more to come. Whittaker is working with an architect who has some initial ideas about weaving paths over the bridge and around the park, connecting the depot as a part of a greater walking trail for Rockdale. He is also working with the Rockdale Parks Board and the historical commission to work on design, usage, and restoration. Also, the park will house a retired calaboose and perhaps a log cabin or two. The Milam County Historical Commission will be seeking National Register listing for all eligible bridges as well as markers from the Texas Historical Commission. It will be a family place with playground equipment and picnic tables. Whittaker said the city is considering moving the Henderson Crossing Bridge, which is currently at the skate park, to join them.

Help for restoring bridges is available from the National Historic Bridge Foundation based in Austin. The foundation is the national advocacy organization for bridge preservation in the United States. Founded in 1998, the group provides technical assistance to local groups seeking to preserve their historic bridges, in addition to serving as a consulting party to bridge projects at the request of state and federal agencies, developing educational programs to promote awareness of historic bridges, and maintaining a clearinghouse for information on the preservation via a website, electronic newsletters, and social media (historicbridgefoundation.com). The executive director of the foundation is Kitty Henderson.

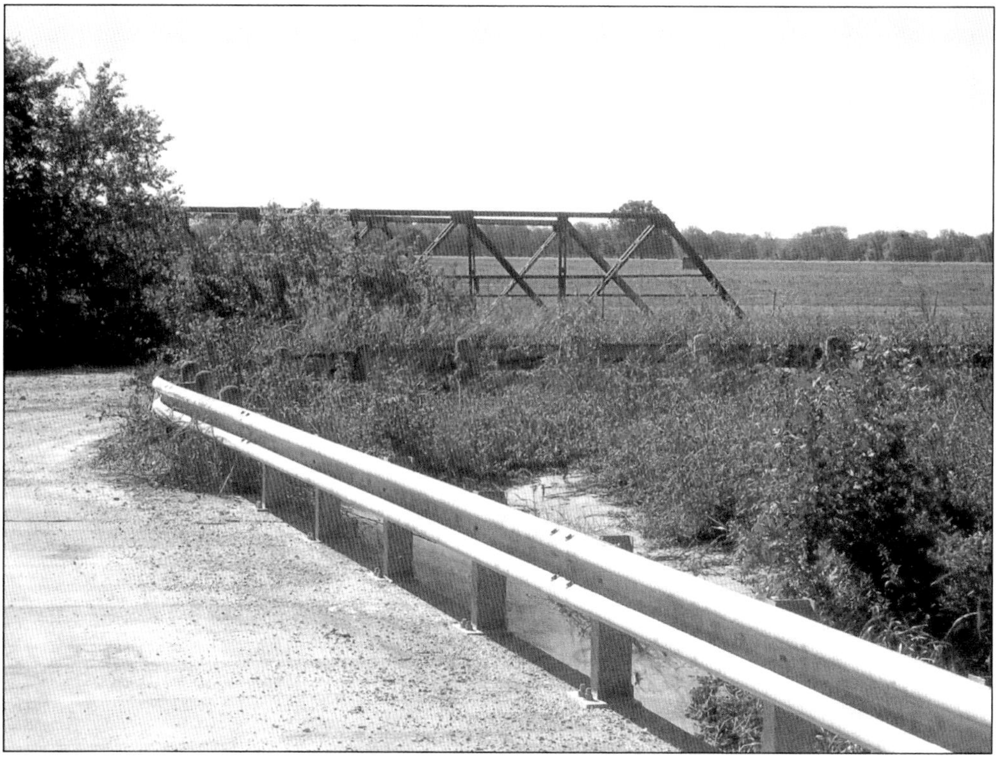

The above photograph of Sheckles Bridge shows it to the south of the new cement slab bridge. County Road 429 crossed the bridge over the San Gabriel River. The photograph below shows Sheckles Bridge sitting in a pasture. This historic bridge was donated to the city by James Perry. (Above, courtesy of David Galbreath; below, courtesy of Jack and Beth Brooks.)

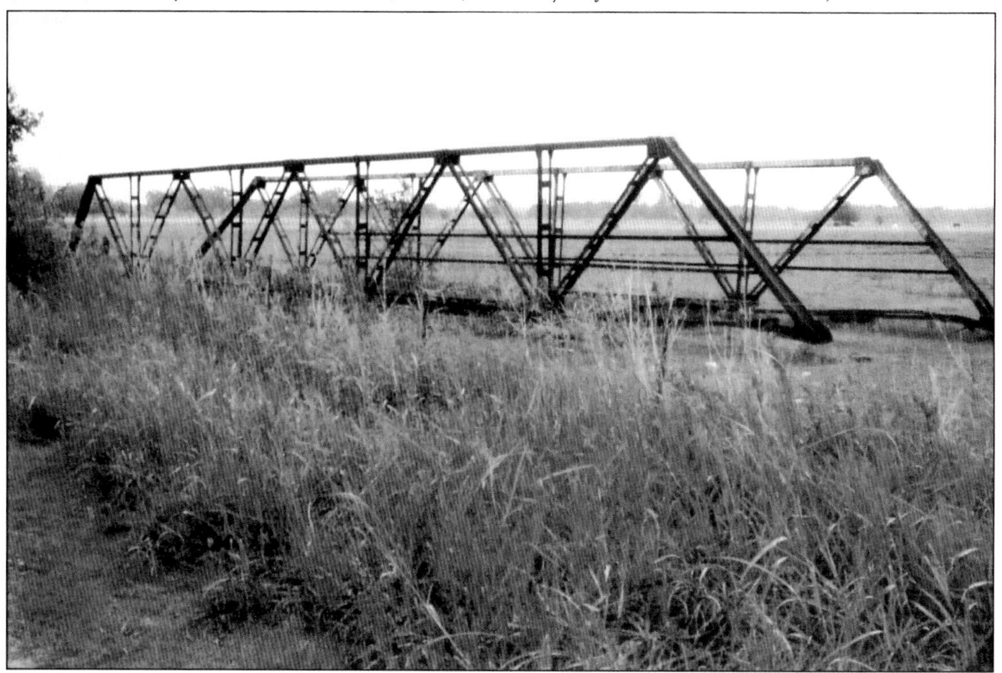

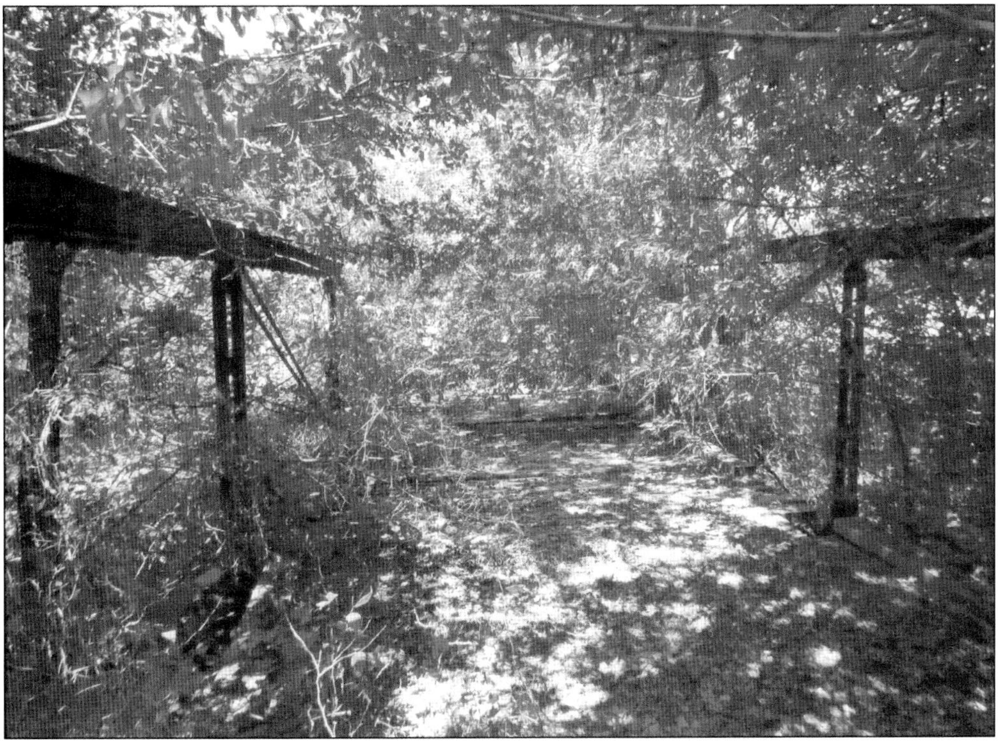

The image above shows Galbreath Crossing Bridge woven into the landscape with years of trees and bushes. The bridge, forgotten and unseen, is covered with leaves. This was taken before it was pulled out of the wooded area. The photograph below shows that a tree had grown through the old wooden deck of the Galbreath Crossing Bridge and had to be cut down prior to moving the bridge to Bridge Park in Rockdale. (Both, courtesy of Jack and Beth Brooks.)

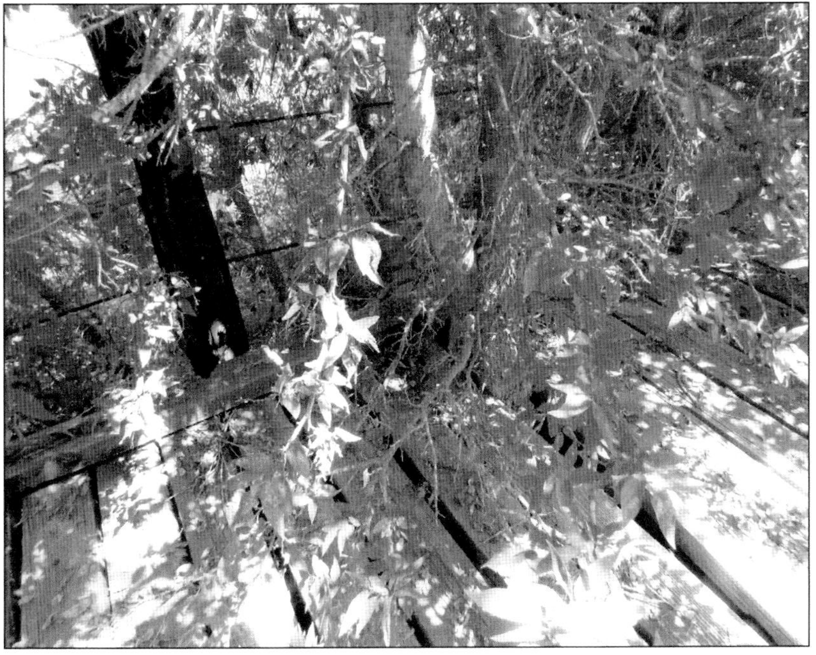

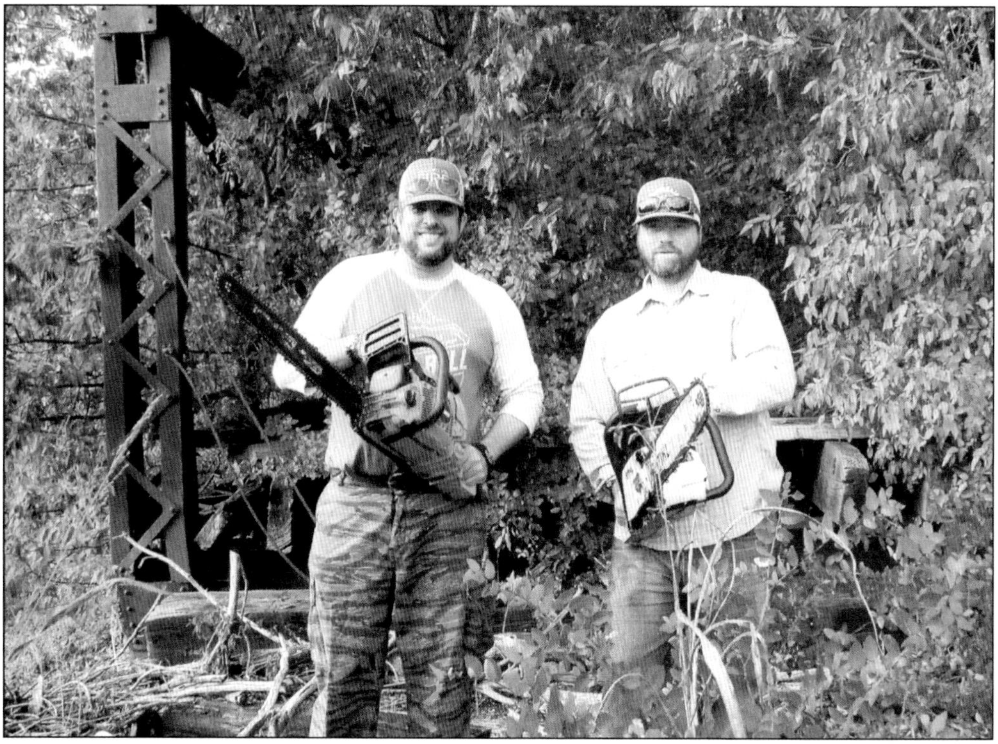

In preparation to start clearing the path to Galbreath Crossing Bridge, workers saw they needed to do some serious tree trimming before the bridge could be moved to Rockdale. Over the years, hackberry trees grew through the decking and intertwined with the trusses, making it challenging work for Rockdale city manager Chris Whittaker (left) and Rockdale Public Works Department director Zack Reeve. These men generously donated their time and have a positive vision for the city of Rockdale and its citizens. (Courtesy of Jack and Beth Brooks.)

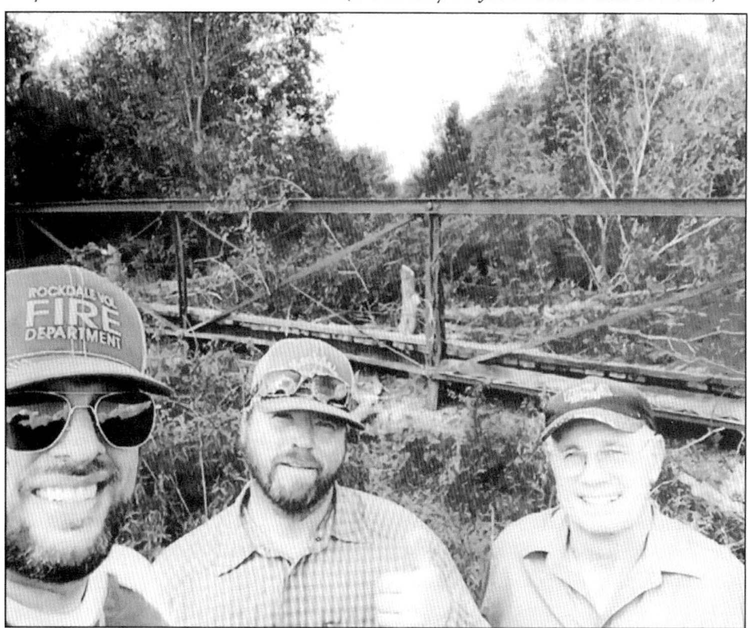

Pictured are, from left to right, Rockdale city manager Chris Whittaker, Rockdale Public Works Department director Zack Reeves, and member of the Milam County Historical Commission Jack Brooks. The citizens of Rockdale give them a thumbs-up for strides made in historic preservation. (Courtesy of Jack and Beth Brooks.)

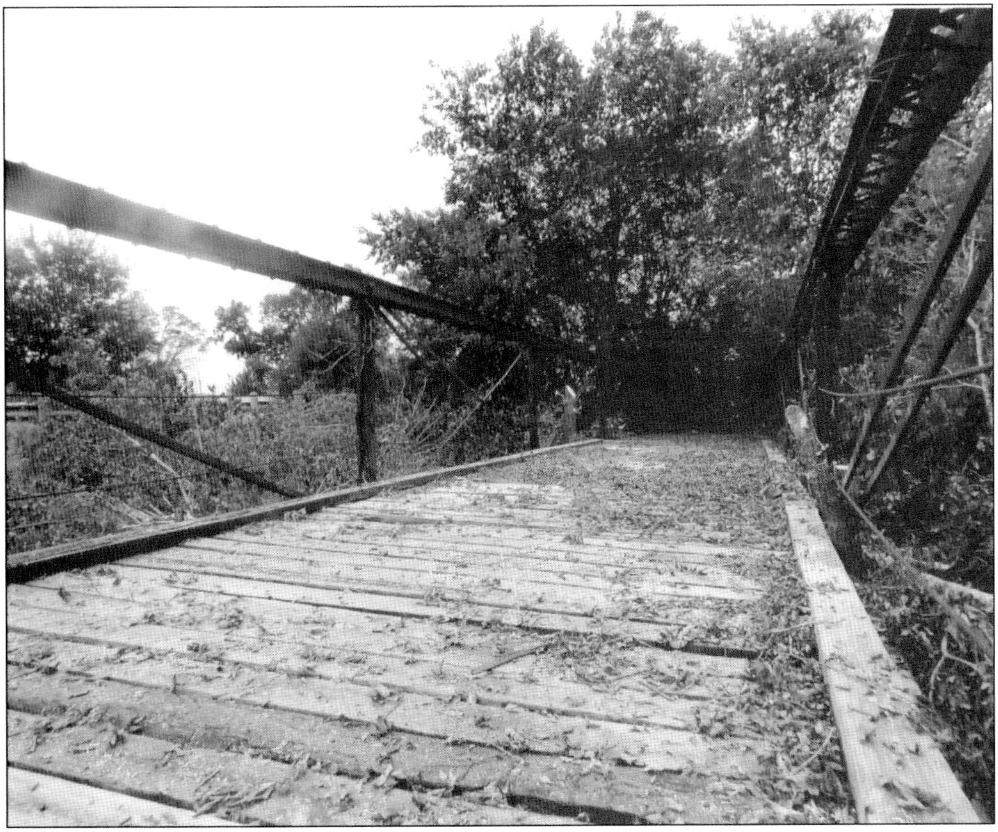

Both photographs show Galbreath Crossing Bridge. Once it was cut free, the bridge was pulled out to the clearing. All of the trees were cut and removed from the bridge deck before it was transported to Rockdale. (Both, courtesy of Jack and Beth Brooks.)

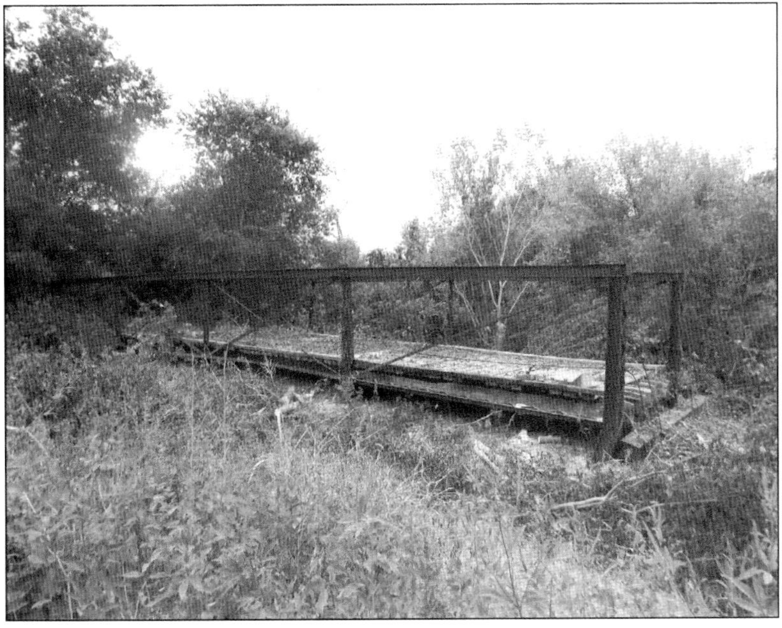

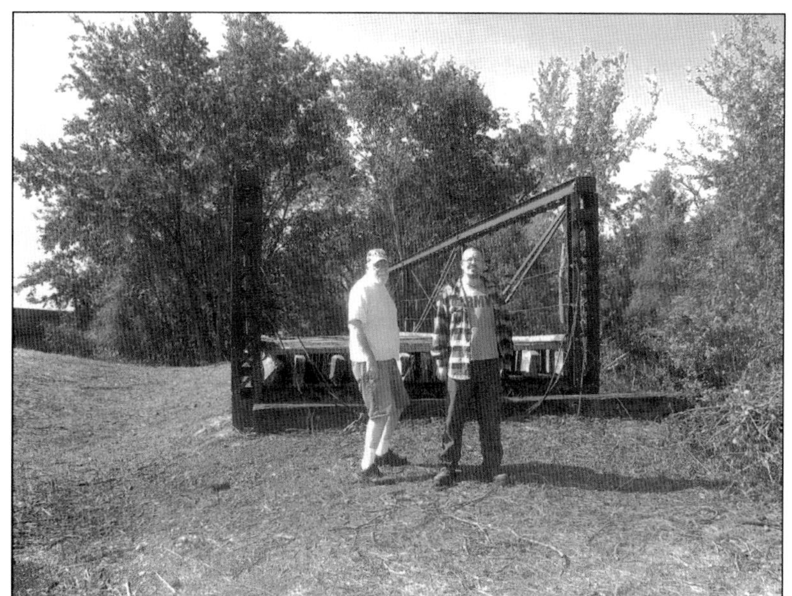

This photograph was taken before the bridge was moved to Rockdale Bridge Park. David Galbreath and his son Derek pose at the Galbreath Crossing Bridge on County Road 240, north of the Cameron Airport at Elm Creek. (Courtesy of David Galbreath.)

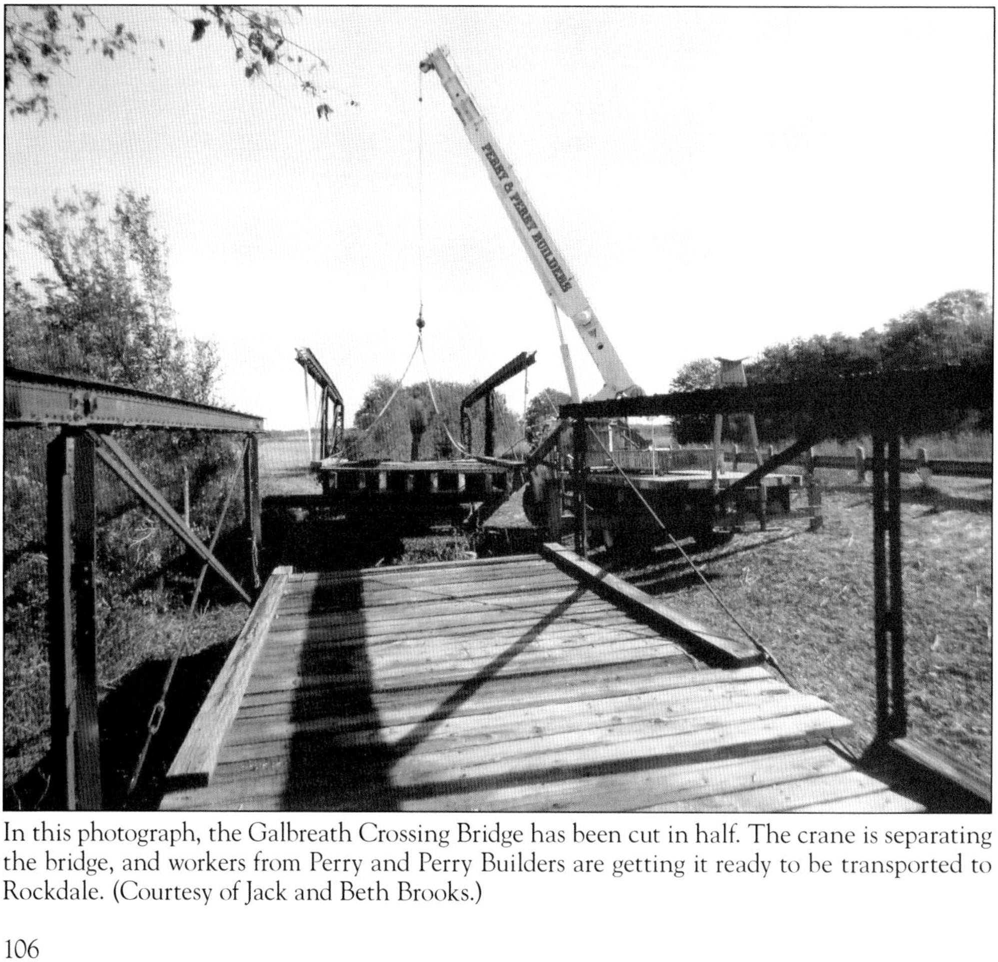

In this photograph, the Galbreath Crossing Bridge has been cut in half. The crane is separating the bridge, and workers from Perry and Perry Builders are getting it ready to be transported to Rockdale. (Courtesy of Jack and Beth Brooks.)

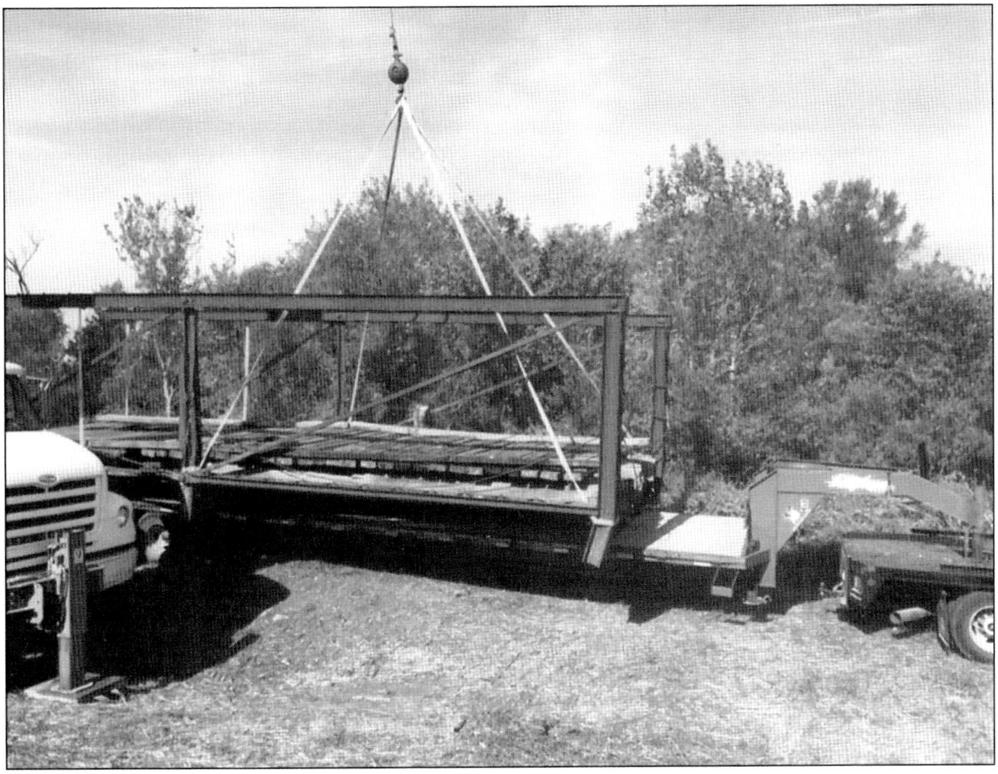

Above, Galbreath Crossing Bridge is being loaded onto a trailer for transportation to Rockdale Bridge Park. The below photograph shows Galbreath Crossing Bridge making its way to its new home at Rockdale Bridge Park with the generous support from Perry & Perry Builders. The company donated equipment, personnel, and time for this project. (Both, courtesy of Jack and Beth Brooks.)

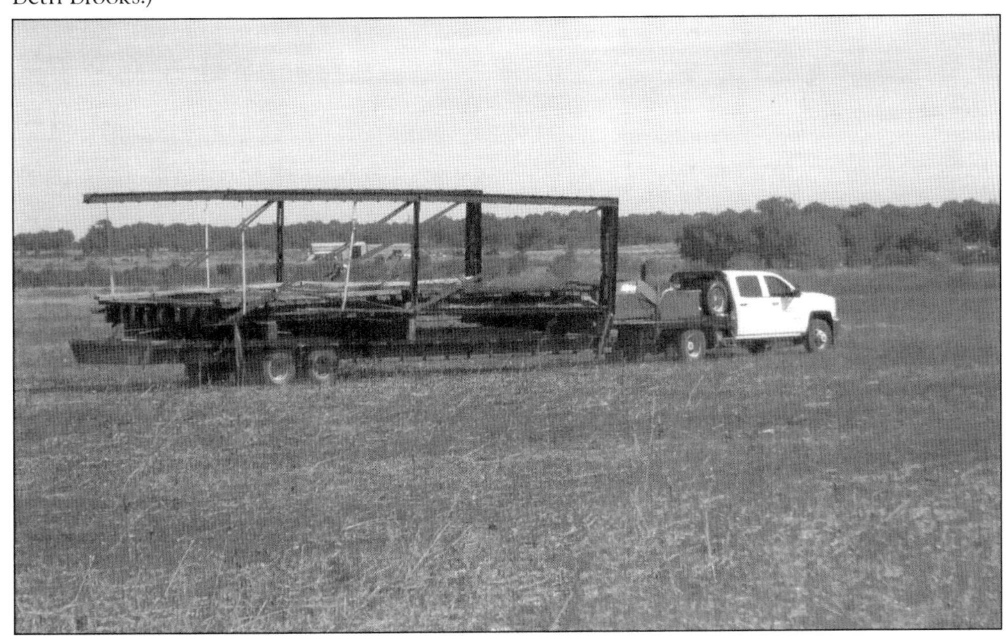

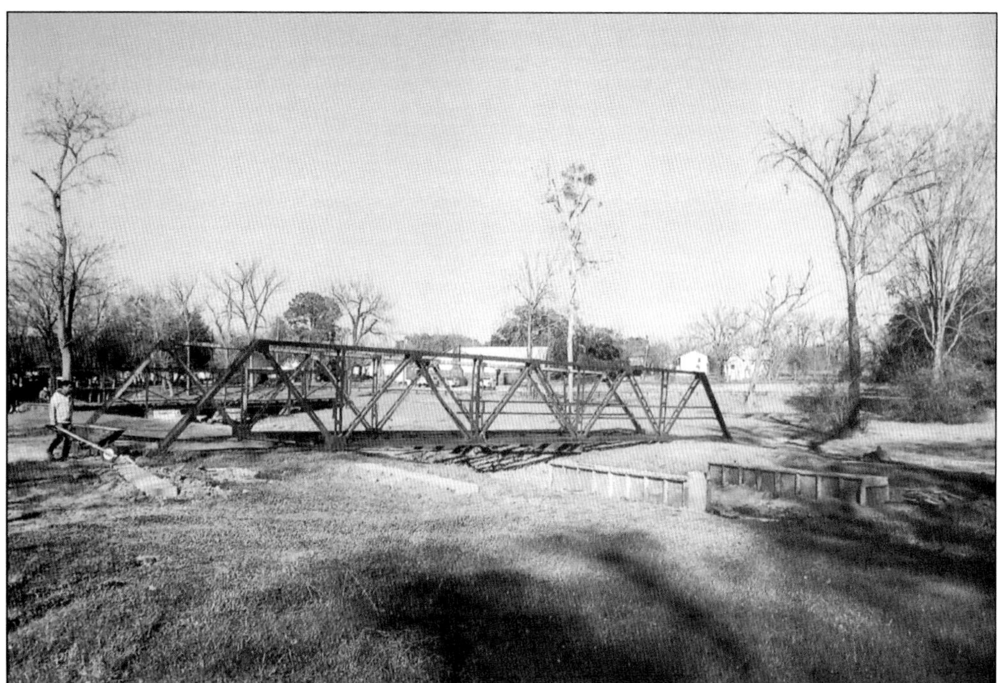

Bridge Park is located on city-owned property in a floodplain located on the south side of US Route 79 and just to the west of Southwest Milam Water Supply. In February 2016, workers poured cement piers to get the historic bridges off the ground and make them more stable. In the image above, Sheckles Bridge and the cement piers are shown. There are plans to place both of the historic bridges on cement piers. The photograph below shows Sheckles Bridge, which was moved from the San Gabriel River in 2015. (Both, courtesy of David Galbreath.)

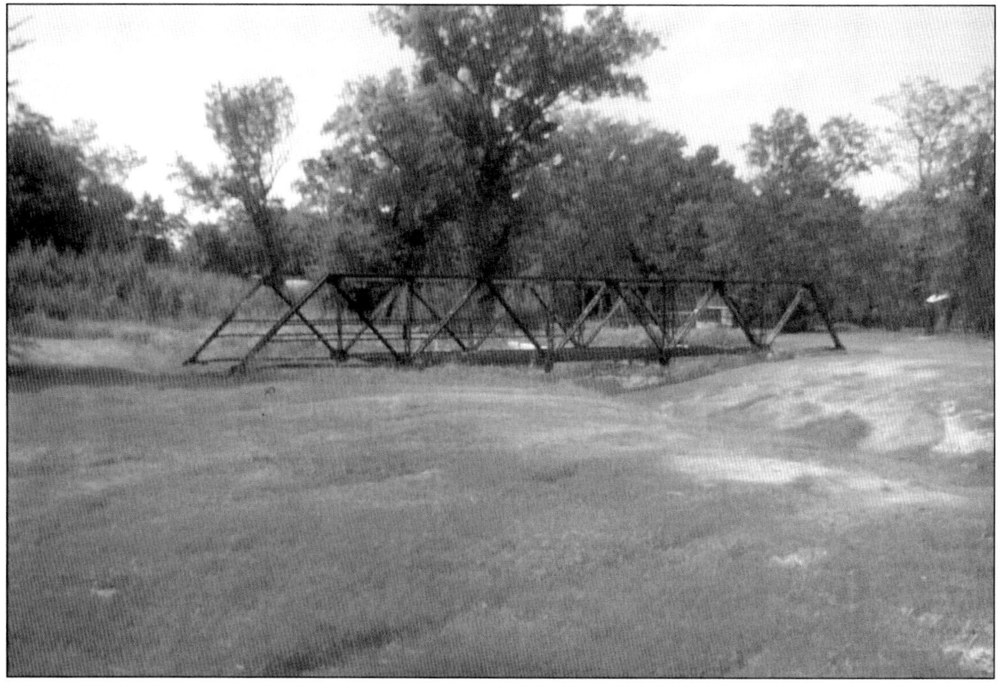

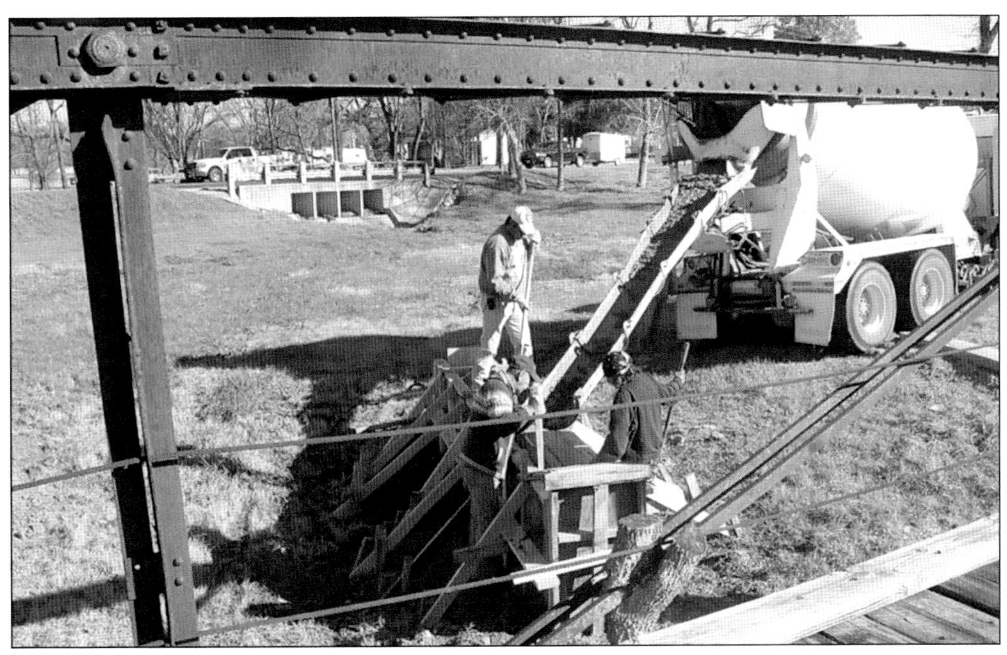

The above photograph is of Galbreath Crossing Bridge waiting be set on the cement piers. Pictured pouring forms for two bridges are, clockwise from lower left, Tommy Doss, Sergio Mora, and Placido Vega. Below is the Galbreath Crossing Bridge after a heavy rainfall, with water running under the historic structure. (Above, courtesy of Mike Brown; below, courtesy of David Galbreath.)

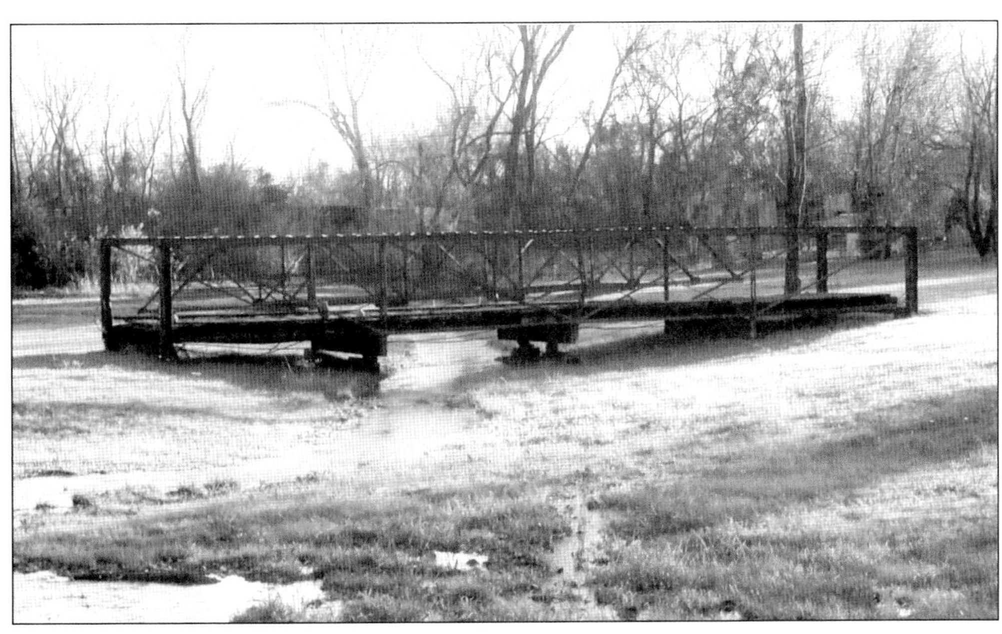

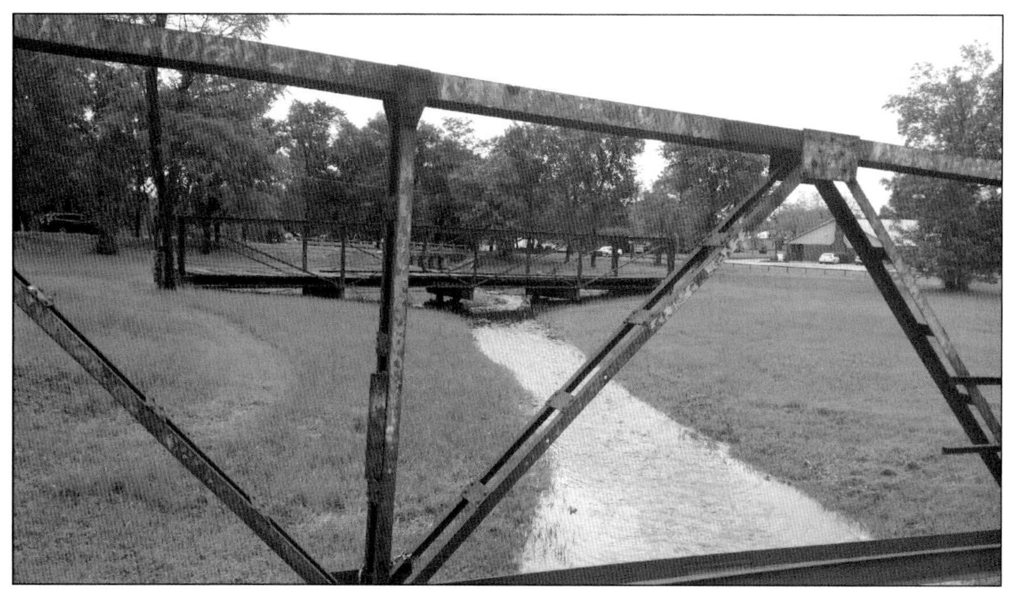

These photographs were taken in August 2016 at Bridge Park after a heavy rainfall. Both bridges have been preserved and repaired, ramps have been built to access the bridges, and benches and trails will be added throughout the park. This will be a great place for local citizens and travelers to stop and visit. People are already migrating here to take pictures or just hang out. (Both, courtesy of Chris Whittaker.)

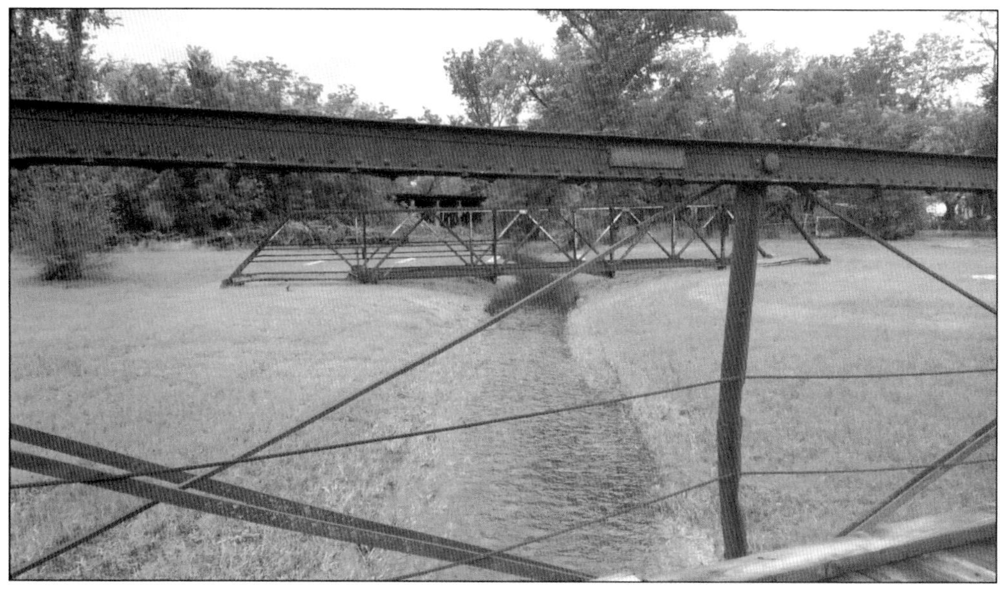

Seven

Renovation of Worley Bridge

Renovation of the Worley Bridge, which is over the San Gabriel River on County Road 428 off Farm to Market 908 near Thorndale, is one of many historic preservation projects that were done to enrich and preserve the deep-rooted history in the county.

In May 2011, county commissioners agreed to work with the Texas Department of Transportation (TxDot) on this project rather than replace the 102-year-old steel truss Worley Bridge.

The Worley Bridge was restored through the federal Highway Bridge Replacement and Rehabilitation Program administered by TxDot at significantly less expense than replacing it. Of the $1.2 million restoration cost, the federal government contributed 80 percent; the state, 10 percent; and the county, 10 percent, through work with matching funds. Restoring the bridge for vehicular traffic was actually less expensive than restoring it as a pedestrian bridge would have been.

On November 13, 2013, a crane moved the historic Worley Bridge off its pedestal and relocated the structure to an adjacent field, where it was restored, repainted, and then returned to its strengthened base.

On Sunday, June 22, 2014, at 2:00 p.m., the rededication ceremony began with the posting of the colors and pledges to the flags. "The Star-Spangled Banner" was sung, and an opening prayer was given. A brief history of the bridge was presented, followed by the ribbon cutting. After the ceremony, all who attended enjoyed refreshments.

Worley Bridge has been deemed eligible for the National Register of Historic Places.

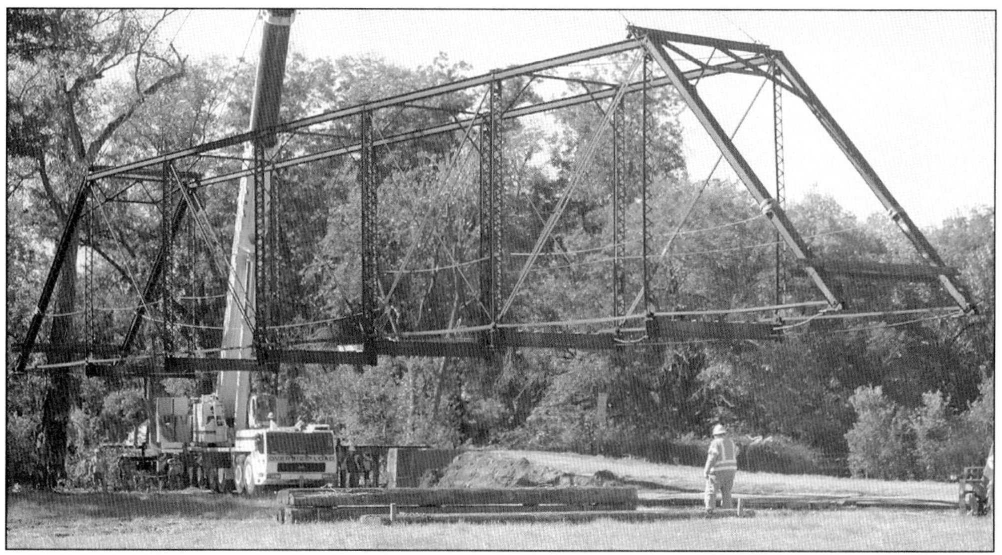

On November 13, 2013, renovation began on the 102-year-old Worley Bridge. Contractors Ellis-McGinnis Construction Company lifted the 34,000-pound bridge off its supports and pilings with a crane and set it in a pasture. (Photograph by Mike Brown, courtesy of *Rockdale Reporter*.)

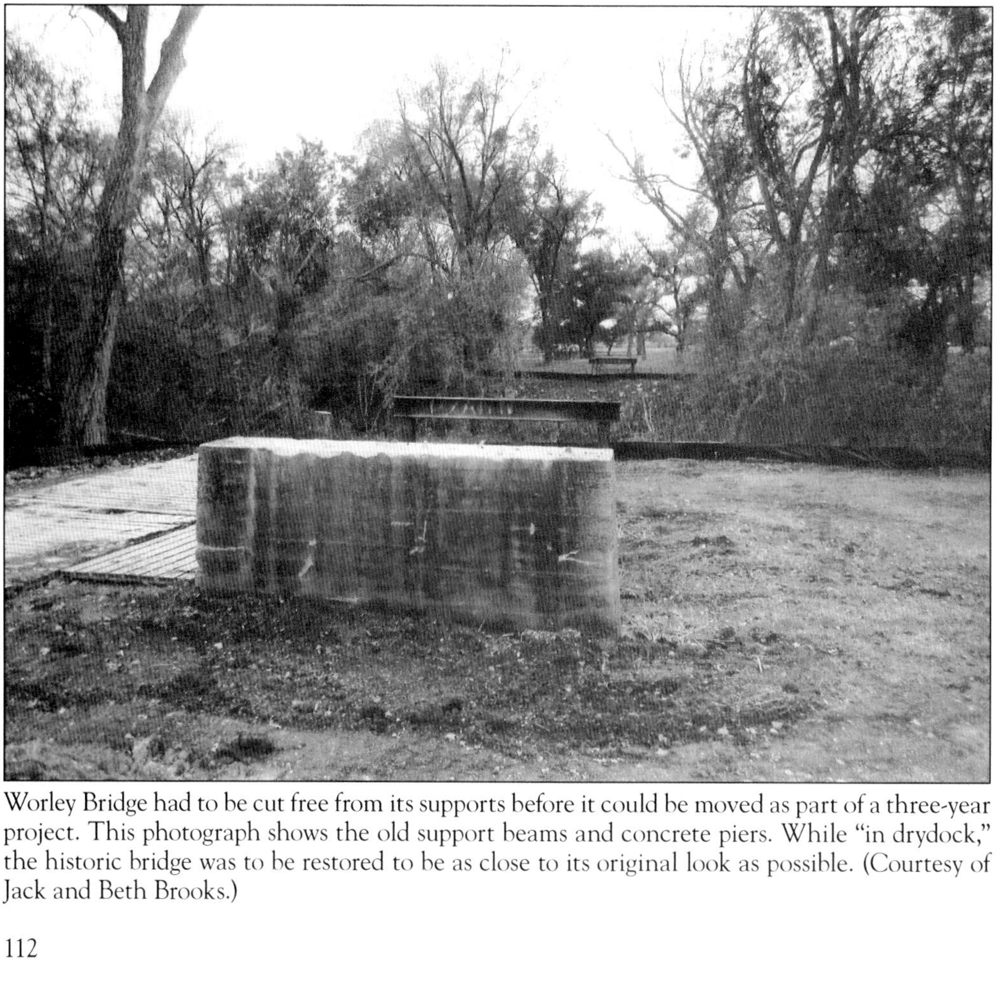

Worley Bridge had to be cut free from its supports before it could be moved as part of a three-year project. This photograph shows the old support beams and concrete piers. While "in drydock," the historic bridge was to be restored to be as close to its original look as possible. (Courtesy of Jack and Beth Brooks.)

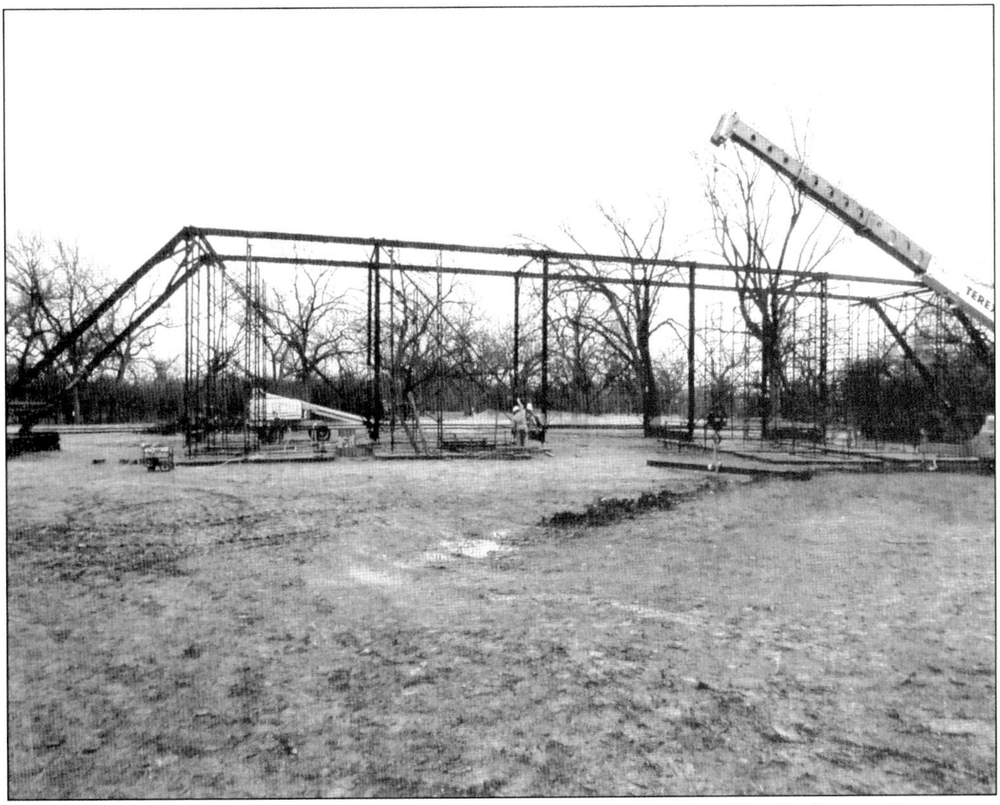

On January 8, 2014, restoration began on the historic Worley Bridge. The decking and stringers were removed from the 136-foot-long bridge in order to lighten the load before it was moved. Workers set supporting scaffolding to place the tent over the bridge. (Both, courtesy of Jack and Beth Brooks.)

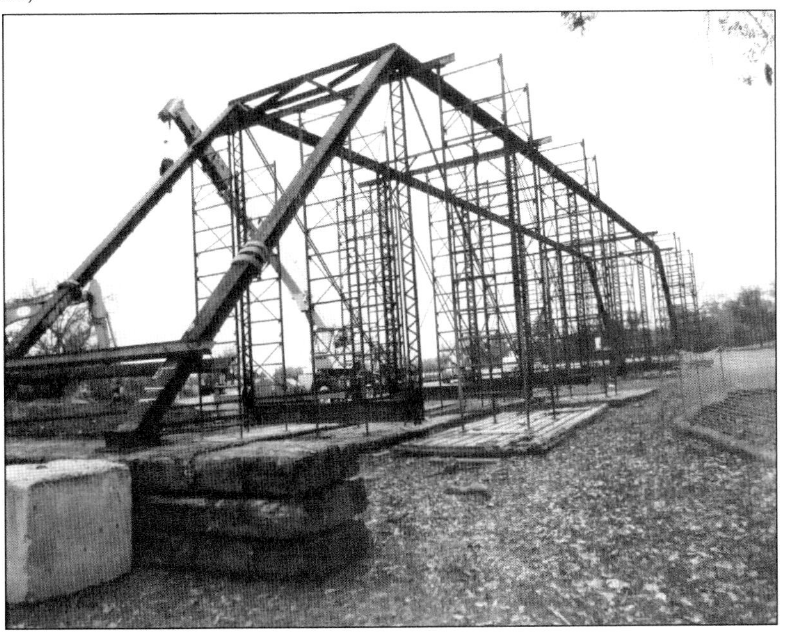

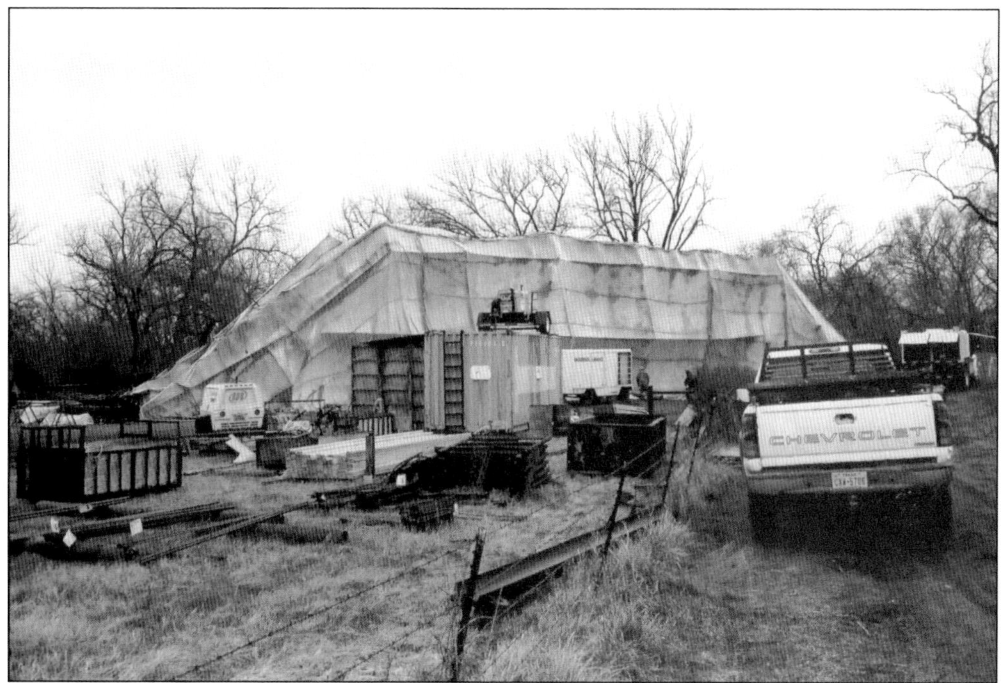

These photographs taken on February 12, 2014, show the tents that were placed over Worley Bridge for the restoration. The Texas Department of Transportation worked with Ellis-McGinnis Company of Eddy. Under the terms of the contract, the company had eight months to complete the project. Other subcontractors moved to the bridge site. (Both, courtesy of Jack and Beth Brooks.)

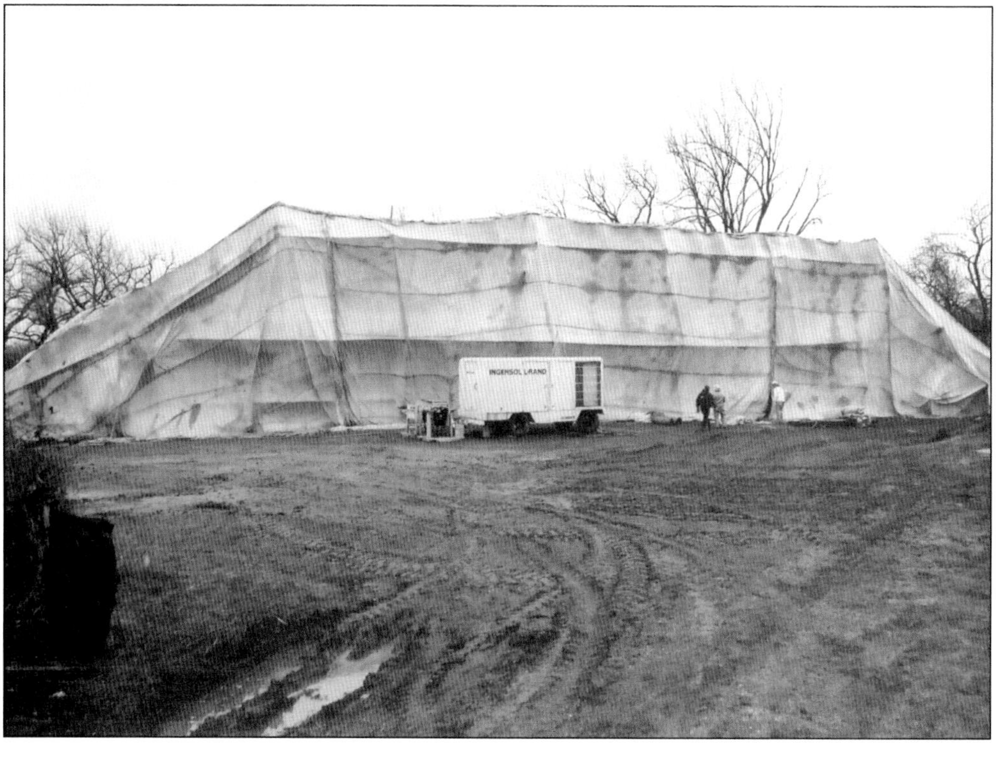

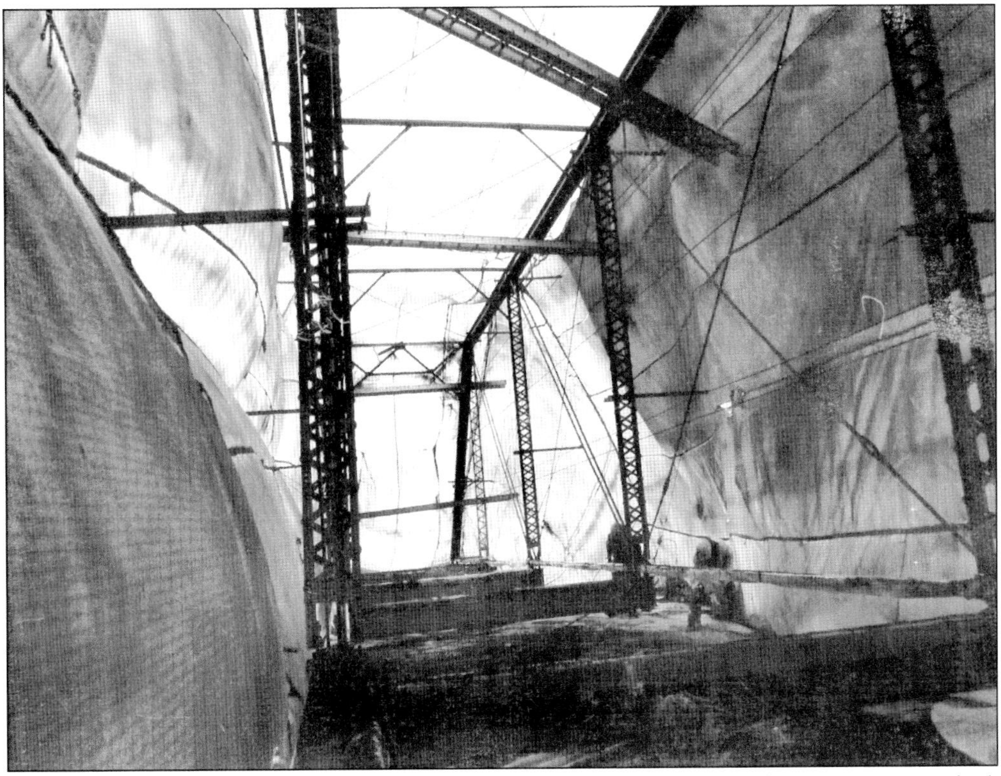

Inside the tent during the renovation, the Worley Bridge was sandblasted, a primer was applied, and the bridge was painted and restored. It would then be returned to the original site. (Both, courtesy of Jack and Beth Brooks.)

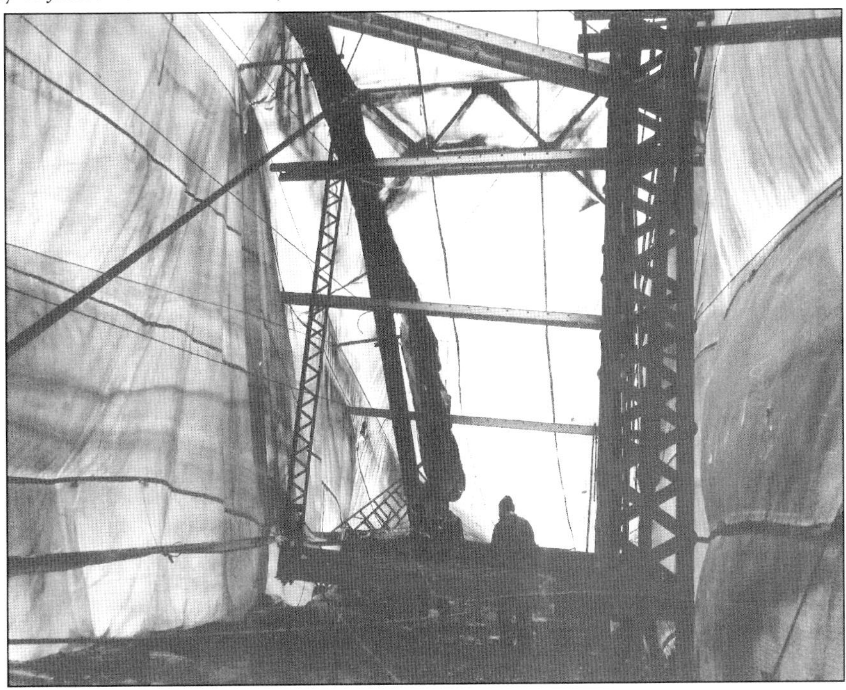

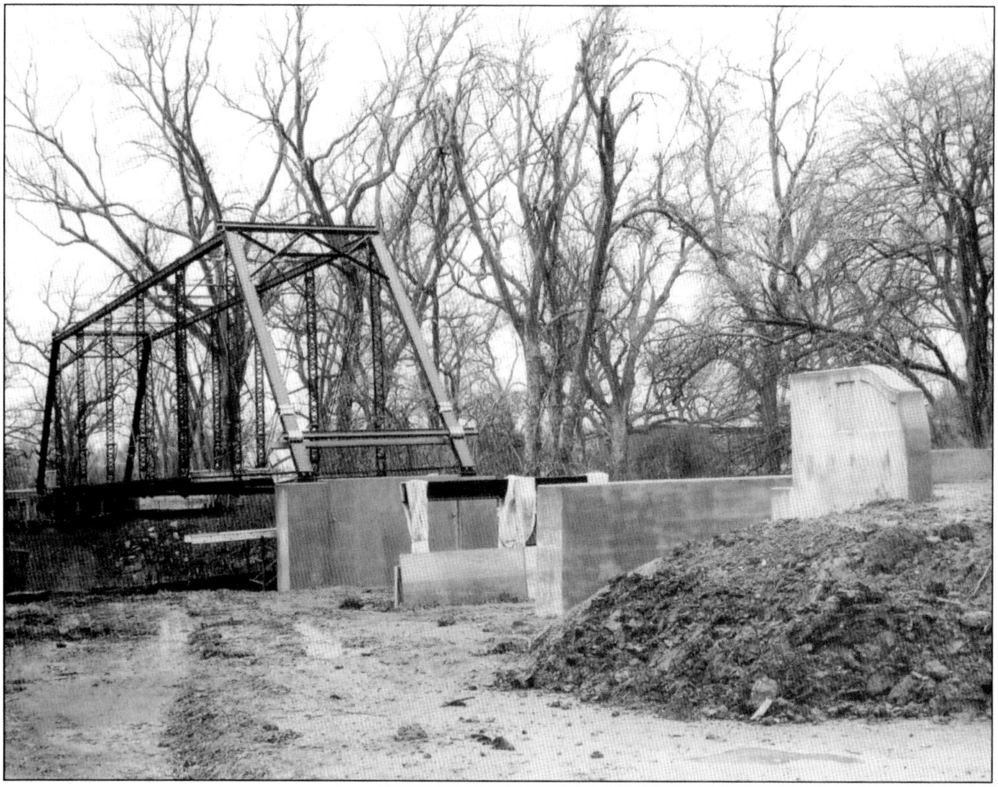

March 2, 2014, was the date that the 103-year-old Worley Bridge was placed back on its support beams. The approach span structure still needs to be added. When it is reopened, it will be for light vehicular traffic. The county road has been closed at the river for almost three years. (Both, courtesy of Jack and Beth Brooks.)

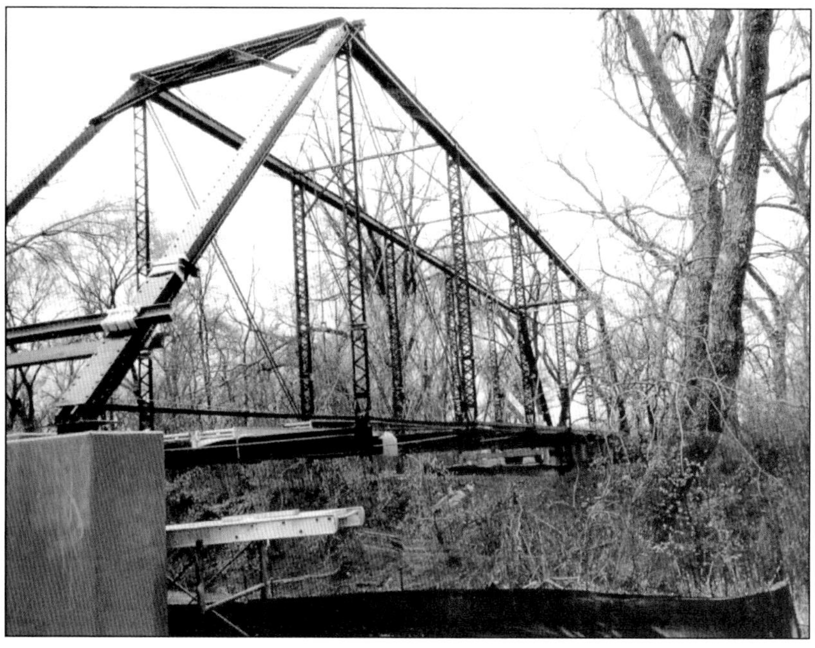

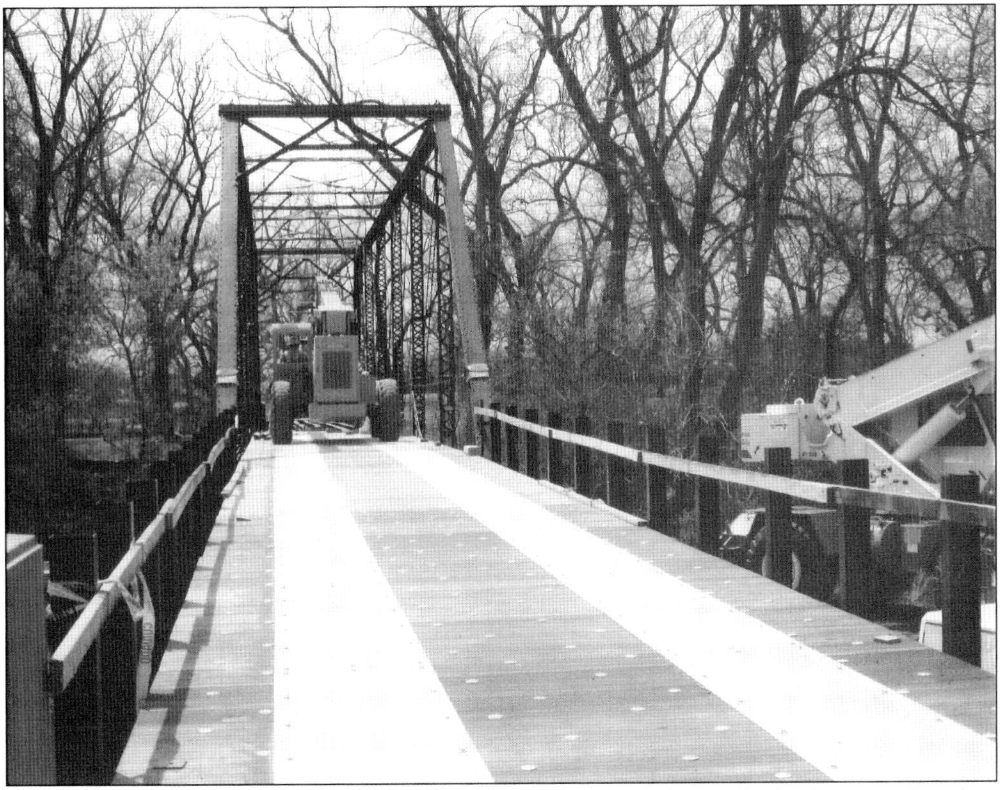

The rehabilitation included replacing the steel grid deck with four-inch glue-laminated timber. These photographs were taken on March 26, 2014. The new decking and stringers were installed after the bridge was returned to its location over the San Gabriel River and placed on new pilings. (Both, courtesy of Jack and Beth Brooks.)

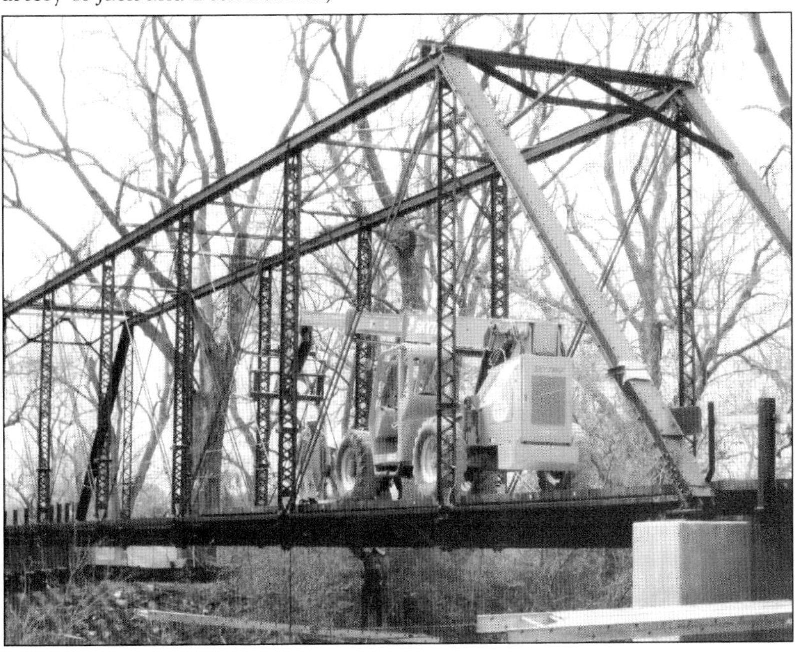

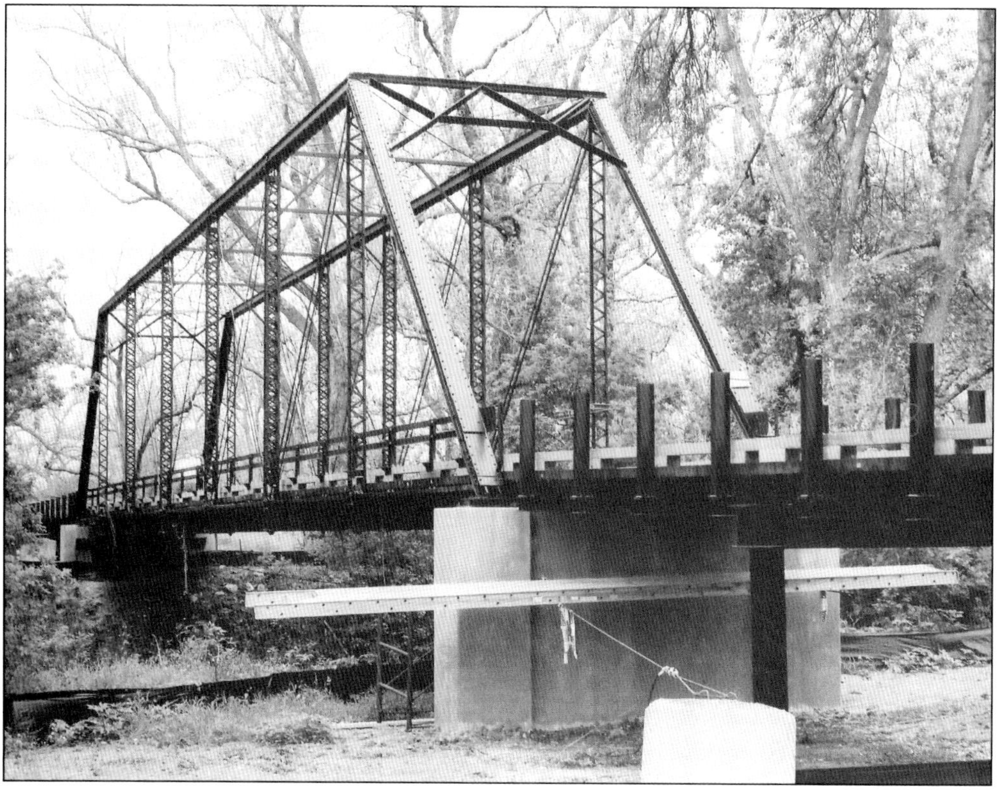

On April 17, 2014, the wood flooring was completed and new concrete piers and bent pier with caisson piers done. The end floor beams were replaced as were all U-bolts and hangers. The plates and nuts, vertical members and end posts, and pedestrian railing were added. (Both, courtesy of Jack and Beth Brooks.)

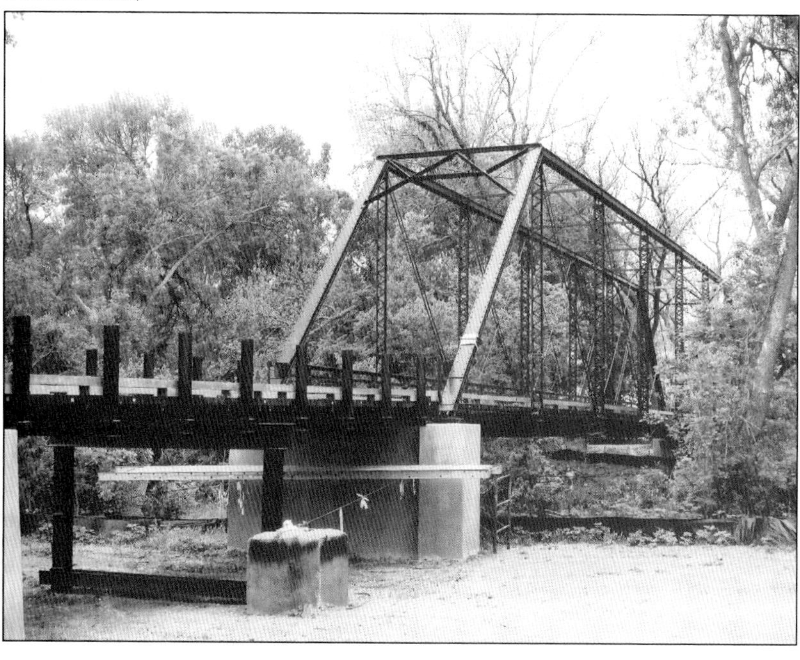

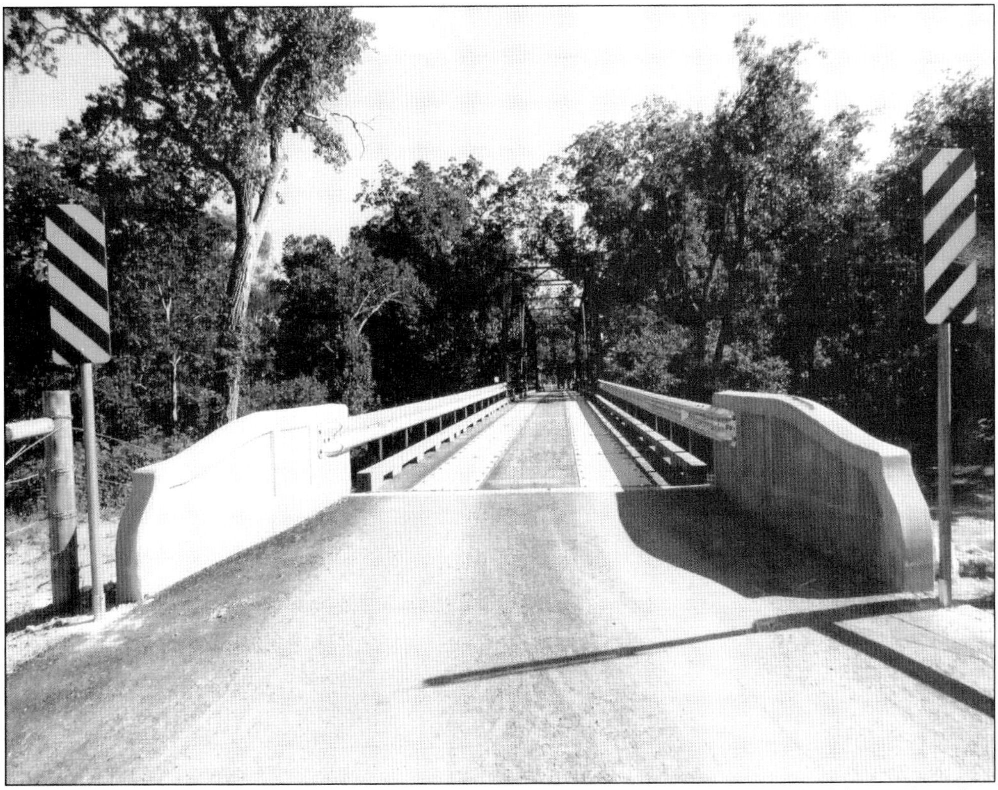

On June 2014, the approach span structure was complete. Worley Bridge had been closed to traffic since renovation began in 2011. It was a wonderful sight to the local residents to have this historic bridge back in operation. During the time the bridge was out of operation, residents solved the problem of not having a bridge by simply fording the river across Apache Pass at low-water crossings. (Both, courtesy of Jack and Beth Brooks.)

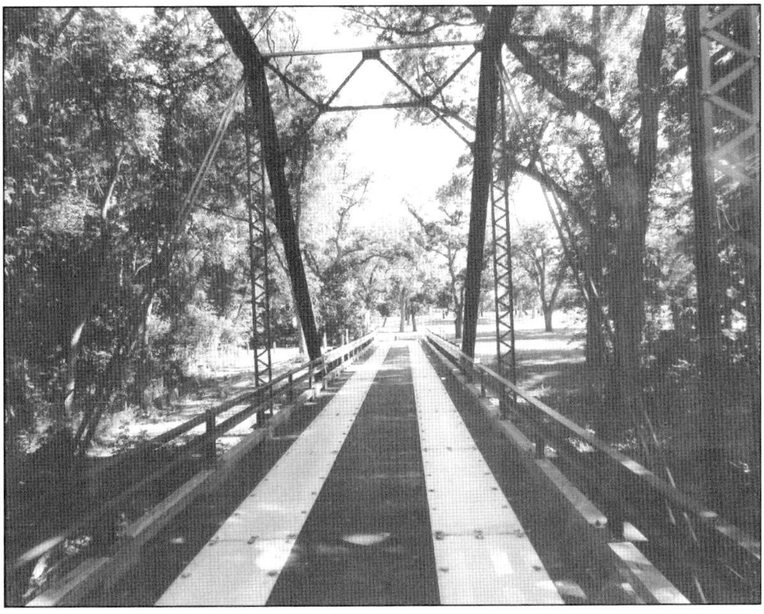

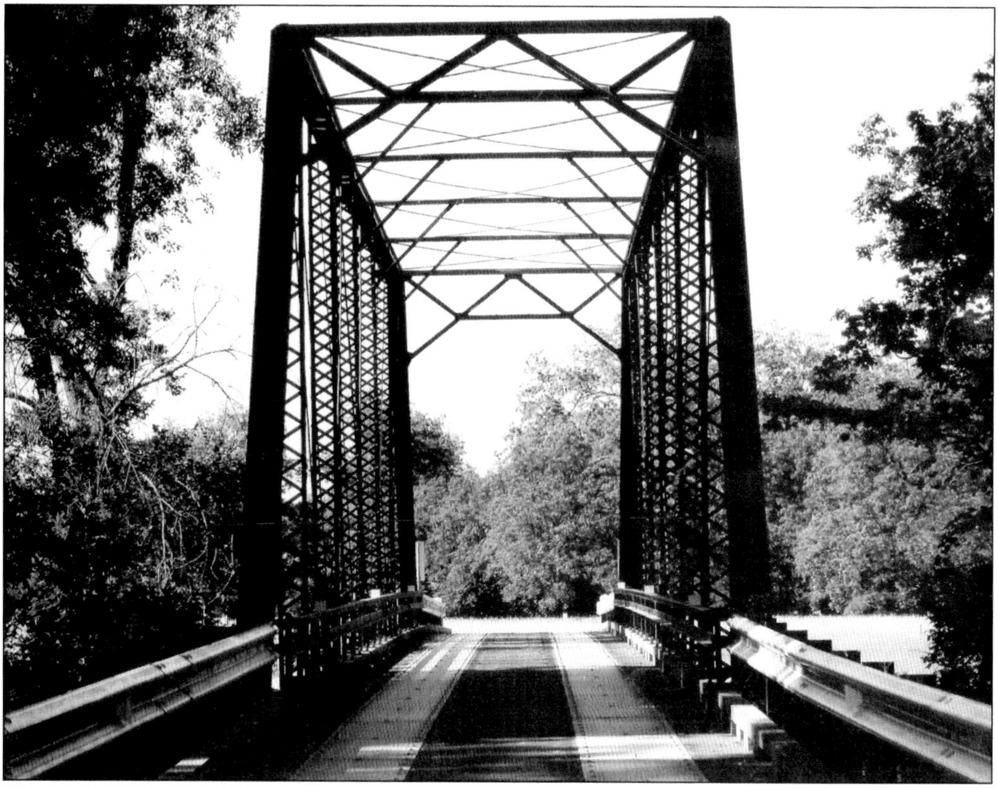

This is the only one of the historic bridges in Milam County that is still used as a traffic bridge—for automobiles and trucks of limited weight. A formal dedication of the bridge was held in 2014. (Both, courtesy of Jack and Beth Brooks.)

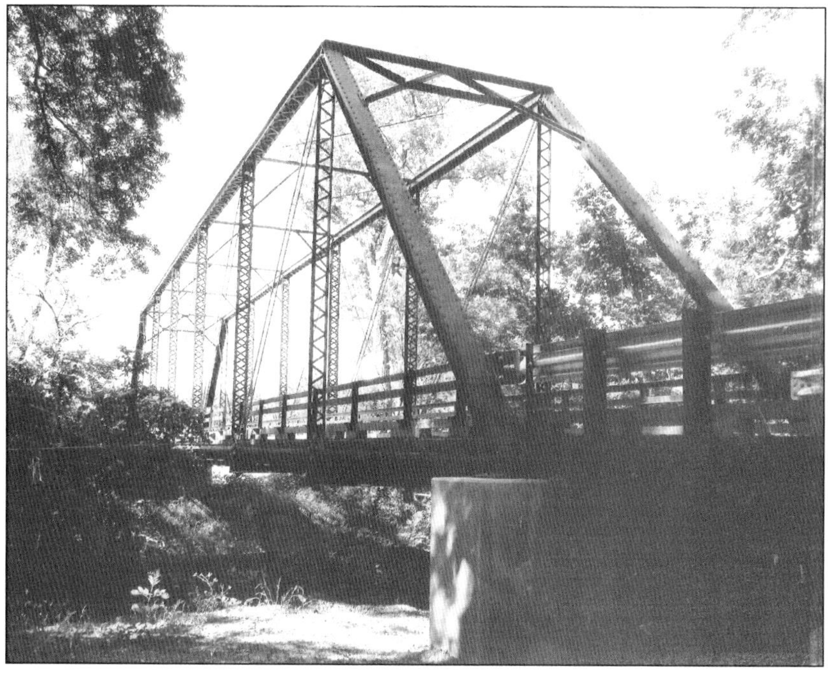

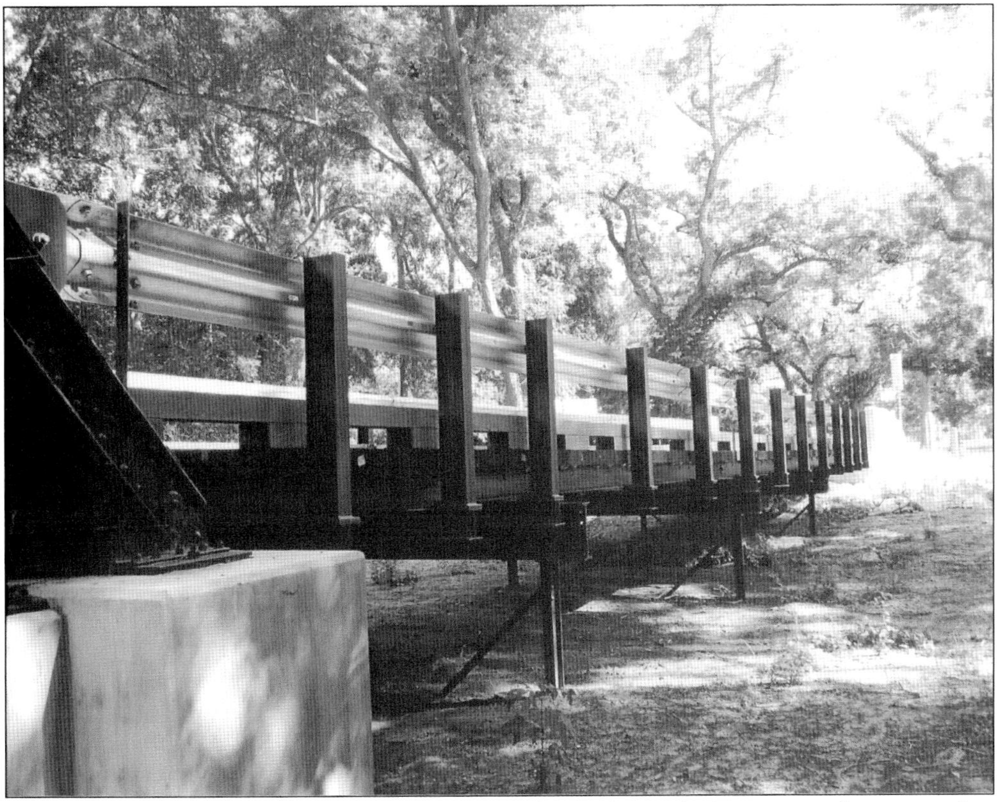

The above photograph shows the side view of Worley Bridge with the new wooden approaches to the historic bridge. The photograph below was taken by someone standing on Worley Bridge and looking down at the flowing water in the San Gabriel River. A new pedestrian railing graces the restored bridge. (Both, courtesy of Jack and Beth Brooks.)

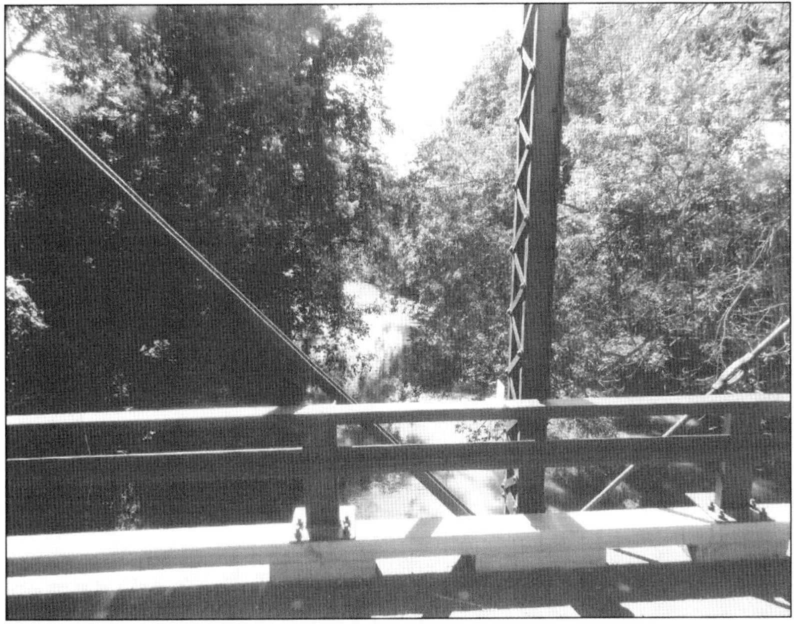

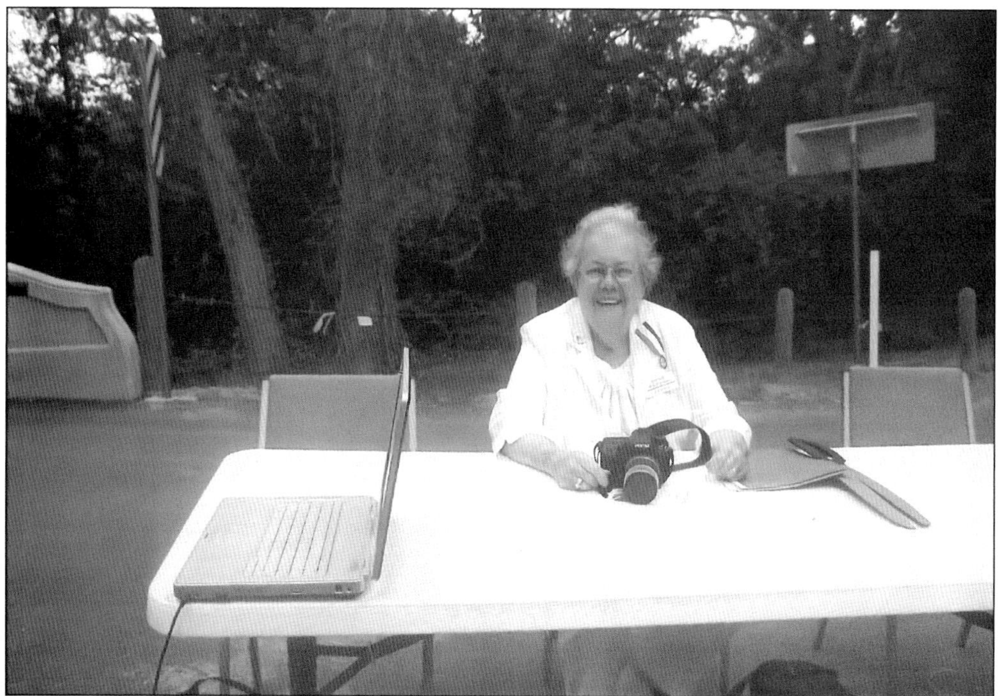

Dr. Lucile Estell spoke at the rededication of Worley Bridge on June 14, 2014. Estell is a former chairperson of the Milam County Historical Commission and a board member of the El Camino Real de los Tejas National Historic Trail Association. (Courtesy of David Galbreath.)

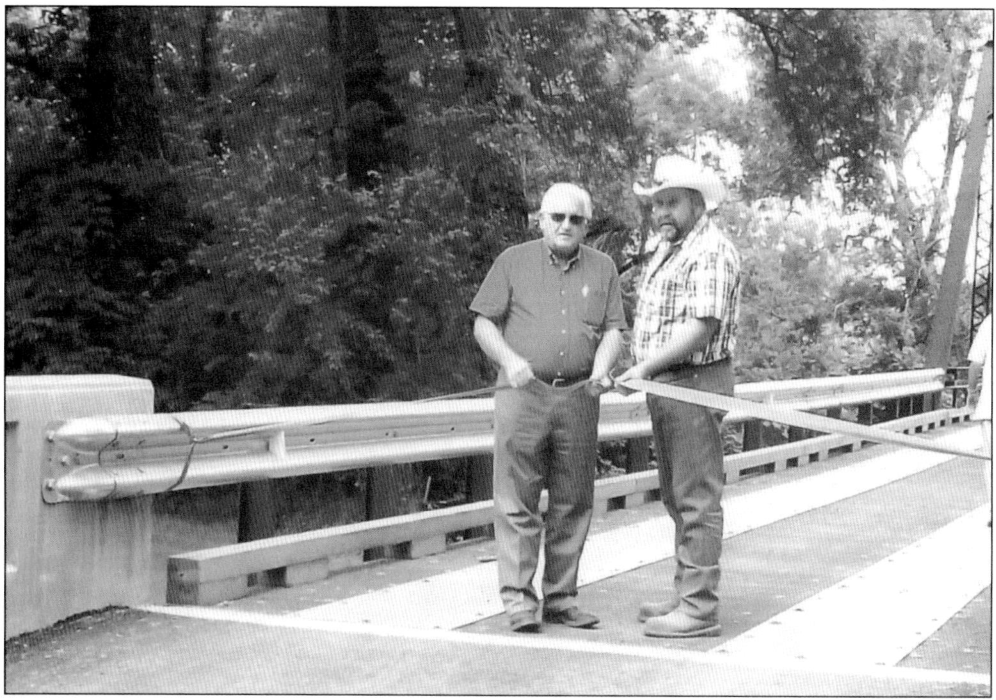

Apache Pass owner Kit Worley (left) and Milam County commissioner Jeff Muegge cut the ribbon at the rededication ceremony. (Courtesy of Joy Graham.)

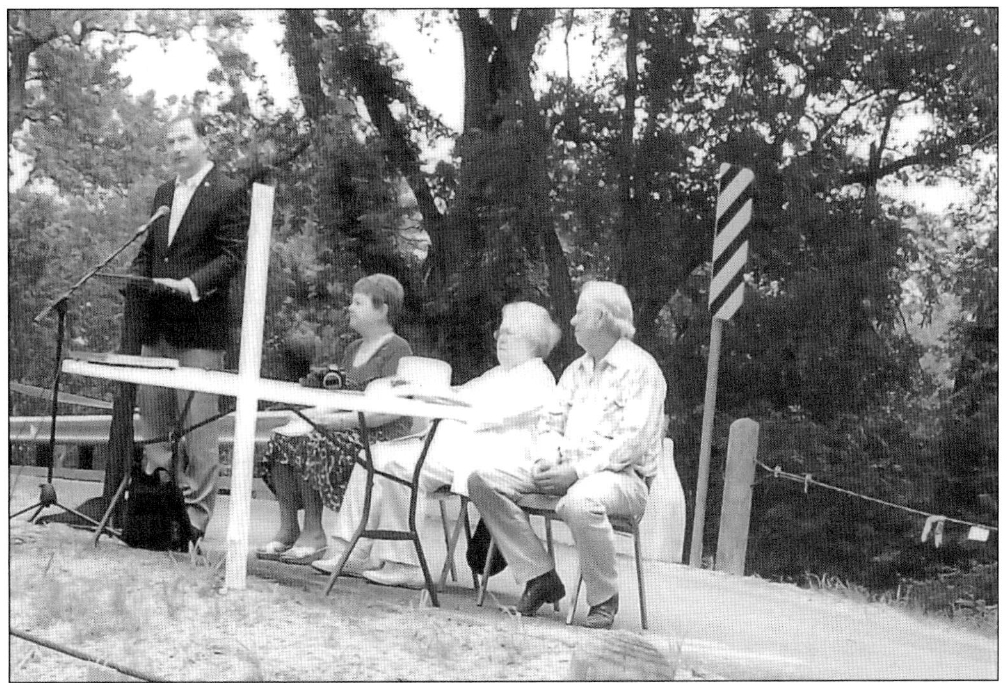

District 5 state senator Charles Schwertner (standing) spoke at the bridge rededication ceremony. Joining him, from left to right, are Geri Burnett, former chair and current vice chair of the Milam County Historical Commission; Dr. Lucile Estell, a member of the Milam County Historical Commission and board member of the El Camino Real de los Tejas National Historic Trail Association; and pastor James Brymer, a member of Milam County Historical Commission, who gave the opening prayer. (Courtesy of Joy Graham.)

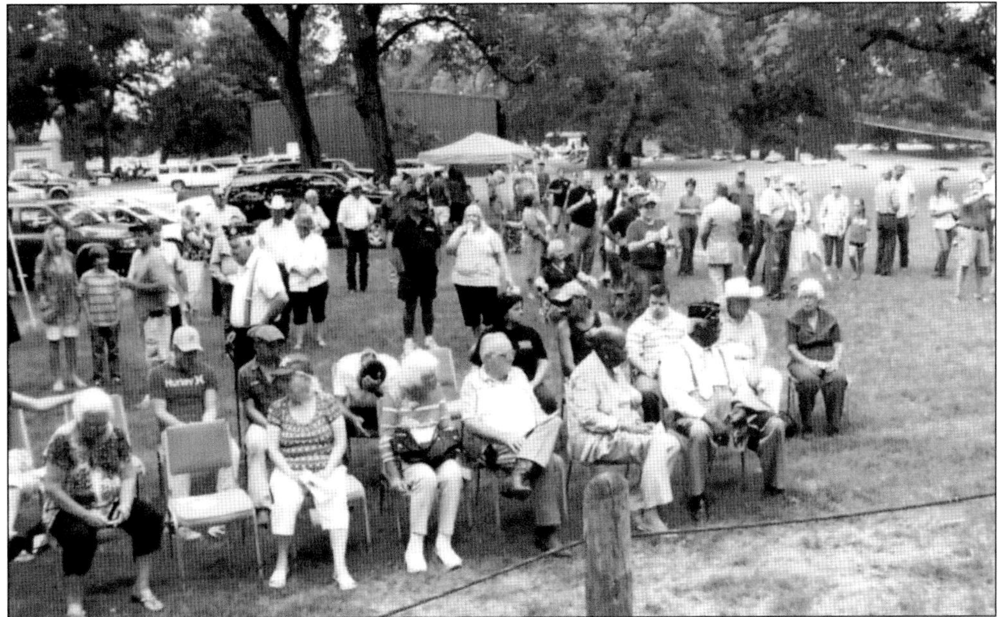

This photograph shows the large crowd that attended the rededication ceremony. After the ceremony, refreshments were served. (Courtesy of Joy Graham.)

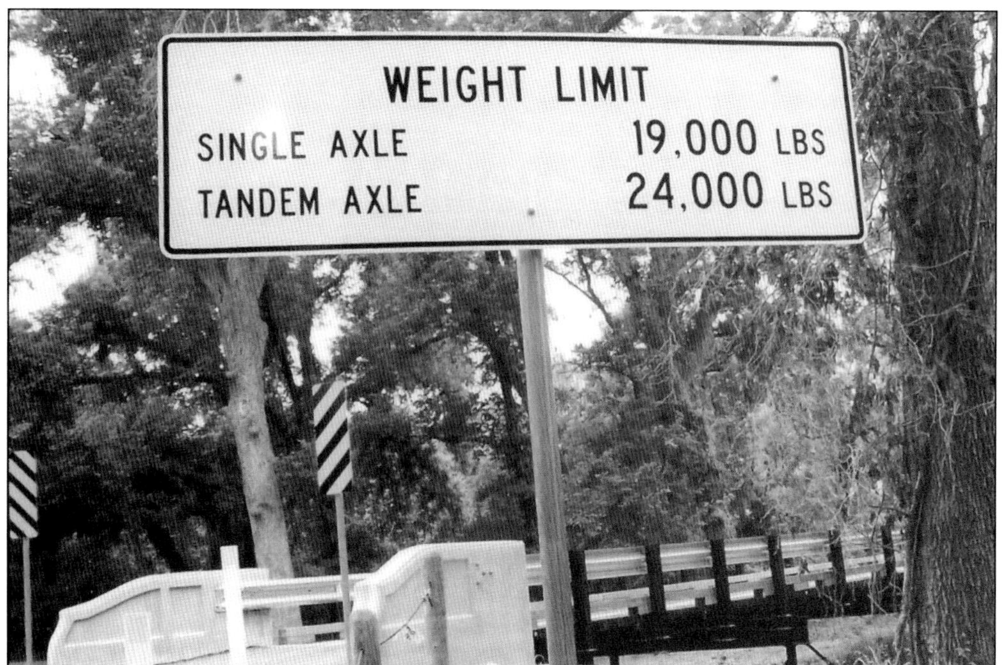

This photograph shows the new weight limit—"Single Axle, 19,000 lbs" and "Tandem Axle, 24,000 lbs." The old weight limit was gross 8,000 pounds and axle or tandem 5,000 pounds. (Courtesy of Joy Graham.)

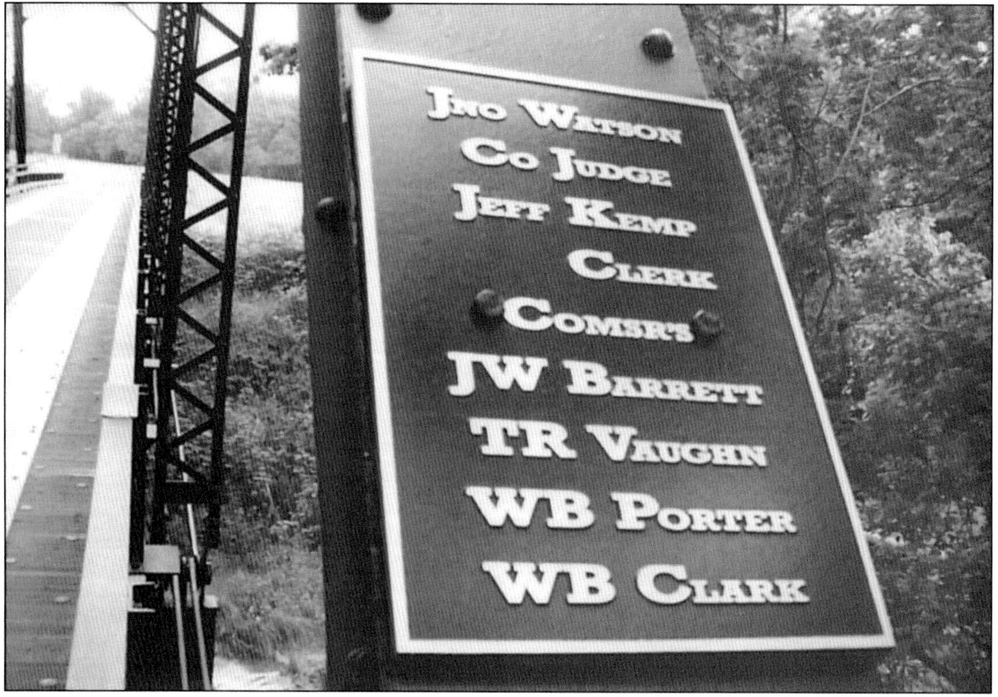

The Worley Bridge plaque that is attached to the bridge lists John Watson as the county judge, Jeff Kemp as the county clerk, and J.W. Barrett, T.R. Vaughn, W.B. Porter, and W.B. Clark as the commissioners. (Courtesy of Joy Graham.)

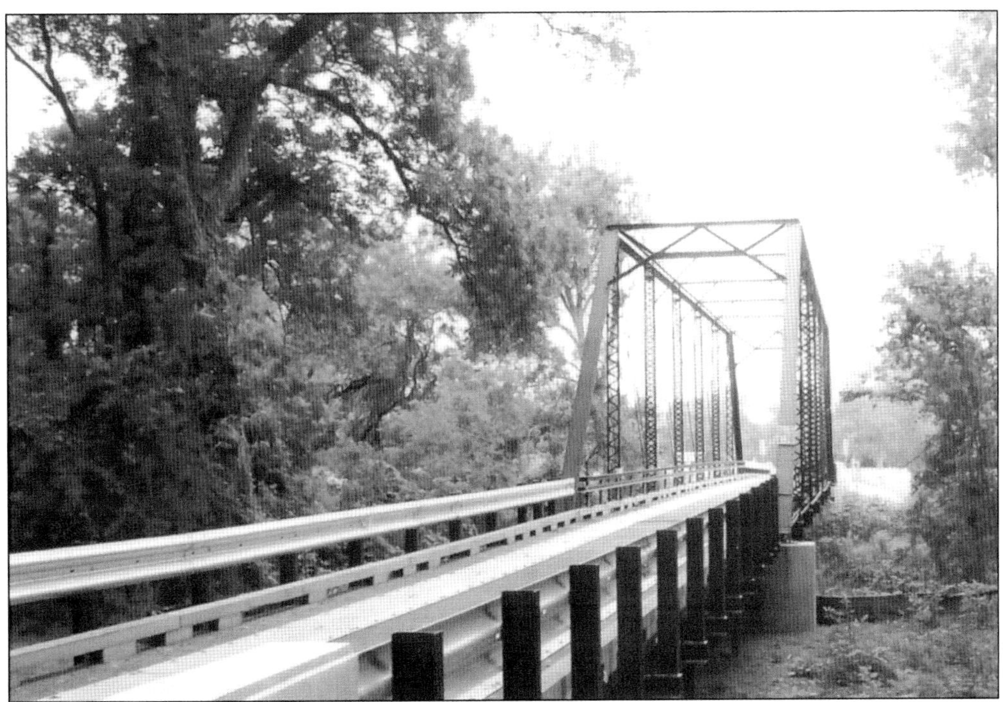

From the side view, one can see the steel construction of this Pratt through pinned bridge that made it tough and enabled it to endure all those years exposed to the weather and other elements. This photograph is of the renovated Worley Bridge. (Courtesy of Joy Graham.)

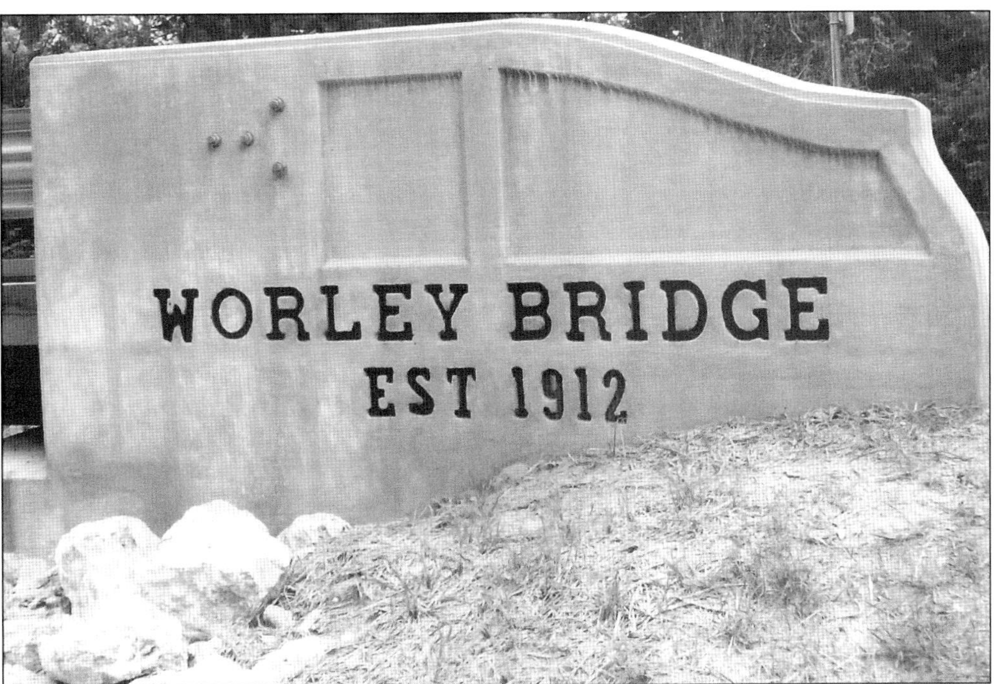

Construction on the Worley Bridge began in 1911 and it opened for use in 1912. The restoration project began in 2012 and was completed in May 2014. (Courtesy of Joy Graham.)

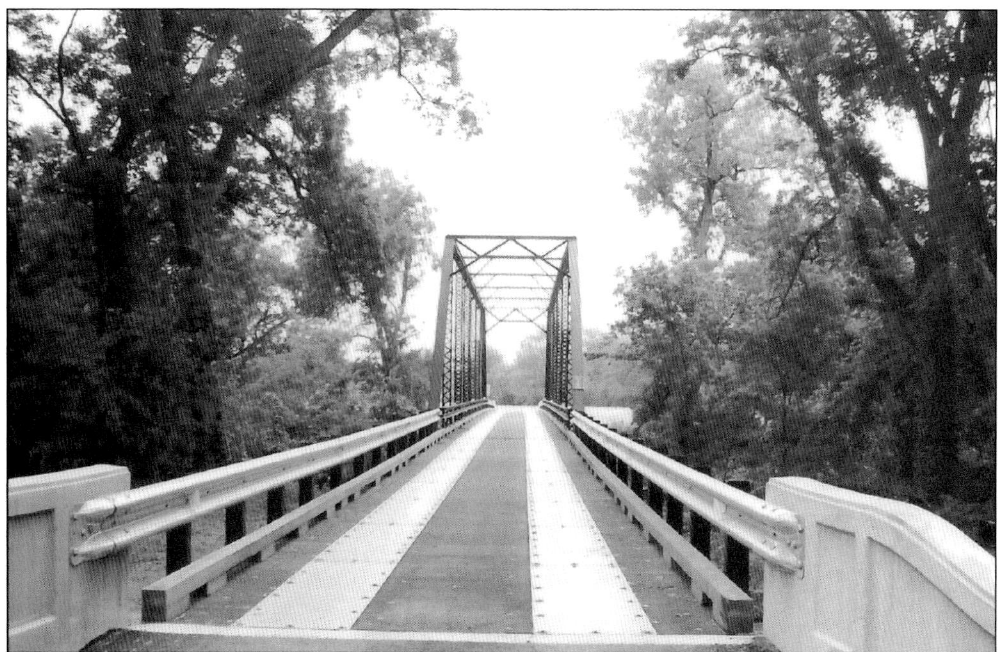

Worley Bridge over the San Gabriel River was first built in the horse-and-buggy days. Since 1912, many generations have crossed this historic bridge. After the restoration, the residents of Milam County remain proud of this historic bridge and look forward to future generations coming and crossing the same path. (Courtesy of Joy Graham.)

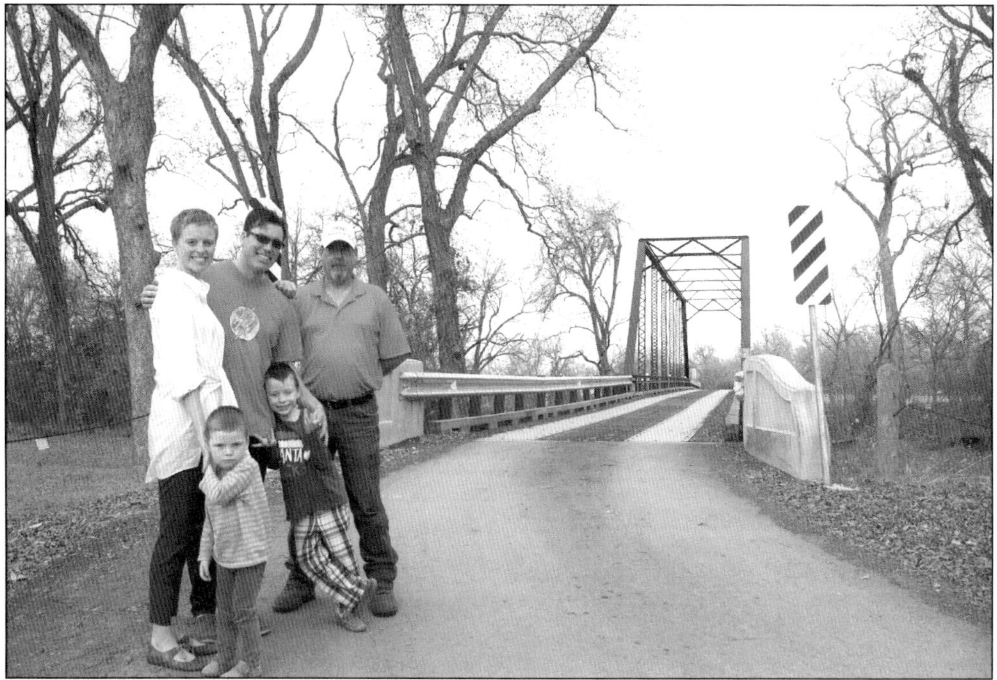

Pictured on a visit to Worley Bridge are, from left to right, Dax, Tiffany, Seph, and Soren Harris from Houston and Mike Harris of Rockdale. The bridge has connected to the Harris family for six generations. (Courtesy of Lucile Estell.)

Musings

David Ray Galbreath was born and raised in Milam County, but since 1984, he had made it his mission to preserve the history of Milam County. His number one passion has been the bridges of Milam County. He has traveled all over the county, researching and documenting the bridges.

While visiting the Lucy Hill Patterson Library in Rockdale in 1985, Galbreath saw a news brochure explaining that Galbreath Crossing was to be bypassed and replaced with a cement slab bridge located just off County Road 240 northeast of Cameron on Big Elm Creek. In 2015, the bridge was moved to Rockdale for the new bridge park.

Learning that there was a bridge with his family name intrigued him to dig for more information. From that time, he has spent endless hours looking at the minutes of the commissioners court for information on roads and bridges. He has created hand-drawn maps documenting the exact location and coordinates. He took photographic evidence of each bridge, even taking along family just to share in each scenic discovery. He has contributed more to the project of documenting the bridges than any other historian.

Born July 15, 1942, to David Russel and Inez Beavers Galbreath, David is the eldest son of four children. On September 5, 1971, while stationed at Barksdale Air Force Base in Louisiana, he has been married to Barbara Hampton for 45 years. They have three children, seven grandchildren, and seven great-grandchildren. After serving 23 years in the US Air Force from 1963 to 1986, he moved to Frankston, Texas. Before retiring, he worked from 1987 to 2012 for the Texas Department of Criminal Justice-Corrections. He is a member of the Milam County Genealogy Society, Texas State Historical Association, the Historic Aviation Memorial Museum in Tyler, and the Association of Missileers.

Pride of his Texas roots and its history has always been a deep love for Galbreath. There is no equal for his passion for Milam County, his family home since 1835. Every historical discovery unearths a lost treasure and story of the lives before us. He always wanted to connect with others about the people and places of Milam County for future generations to admire so they are not forgotten.

He is someone remembered and valued by all who know him. I know him well; he is not just my father, but also my hero.

—Tracy Lynn Galbreath-Young

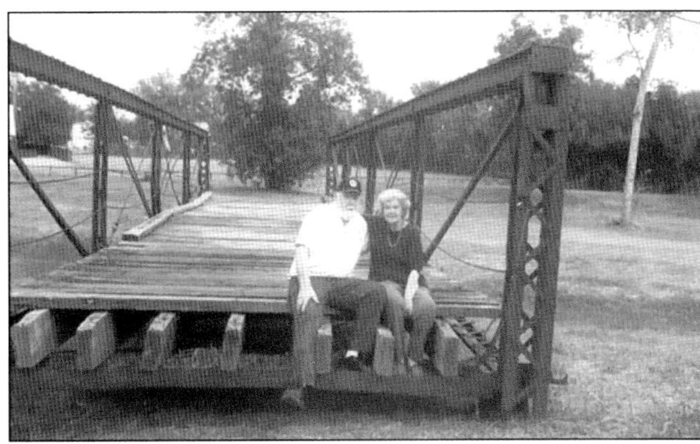

David Galbreath and his mother, Tena Galbreath, of Rockdale, sit on the Galbreath Crossing Bridge in Rockdale Bridge Park off US Route 79. (Courtesy of Tracy Galbreath-Young.)